一部建院史

半座苏州城

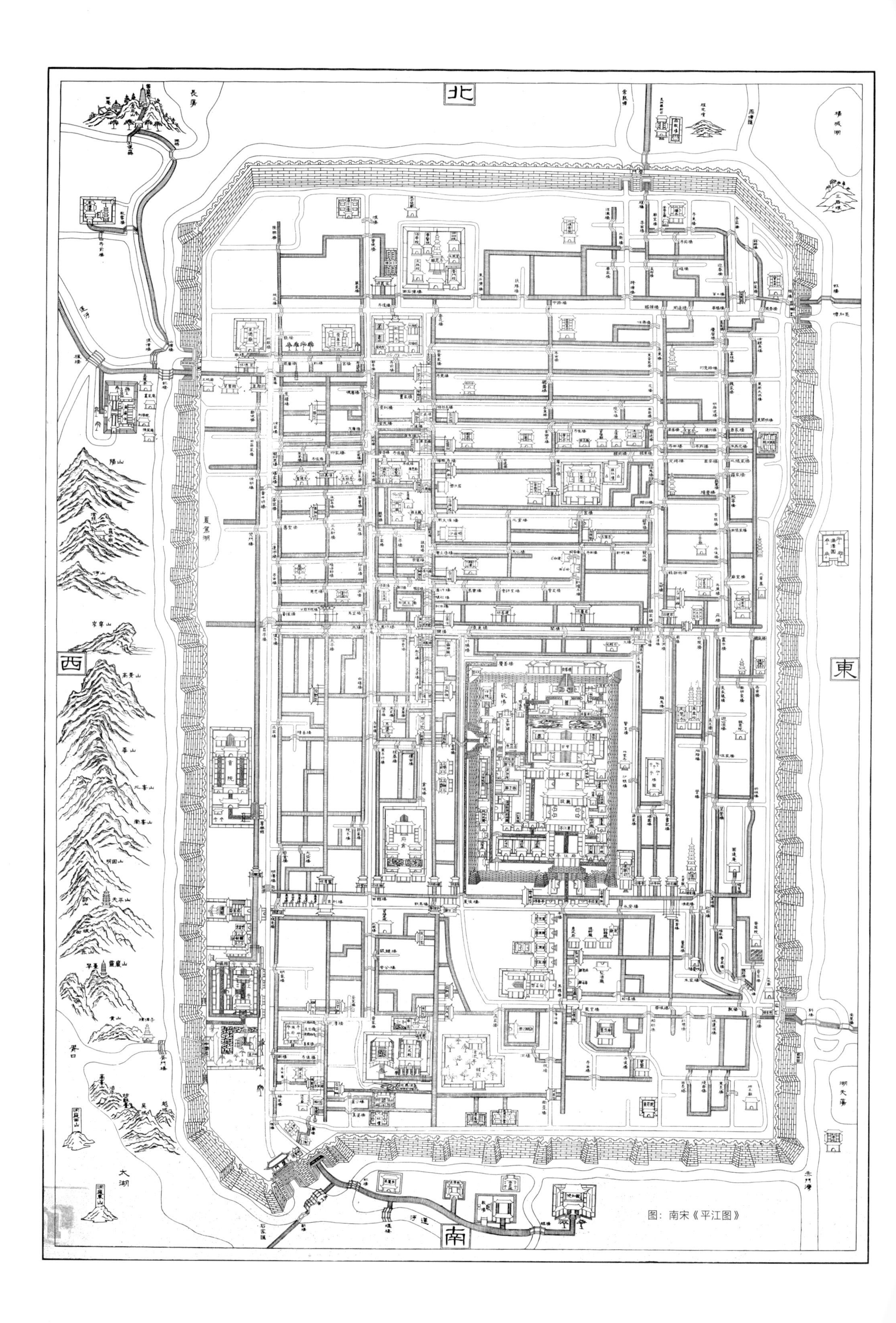

图：南宋《平江图》

启迪设计集团作品集
1953—2023

启迪设计集团股份有限公司　　著

中国建筑工业出版社

图书在版编目（CIP）数据

启迪设计集团作品集：1953-2023 / 启迪设计集团股份有限公司著. —北京：中国建筑工业出版社，2023.9

ISBN 978-7-112-28884-7

Ⅰ. ①启… Ⅱ. ①启… Ⅲ. ①建筑设计—作品集—中国—现代 Ⅳ. ①TU206

中国国家版本馆CIP数据核字（2023）第119746号

责任编辑：丁洪良
责任校对：王　烨

启迪设计集团作品集1953—2023
启迪设计集团股份有限公司　著

*

中国建筑工业出版社出版、发行（北京海淀三里河路9号）
各地新华书店、建筑书店经销
北京锋尚制版有限公司制版
临西县阅读时光印刷有限公司印刷

*

开本：880毫米×1230毫米　1/16　印张：24¾　字数：680千字
2023年10月第一版　　2023年10月第一次印刷
定价：**298.00**元
ISBN 978-7-112-28884-7
（41612）

序

苏州是一座驰名中外的历史文化古城。

春秋时期，吴王阖闾命伍子胥“相土尝水，象天法地”，建成“阖闾大城”，即苏州城，至今已有2500余年。苏州历史积淀深厚，苏州随处可见的历代园林、寺庙、宝塔、桥梁、官衙、名人故居等积淀成珍贵的历史文化遗产，在中国城市建设史书中留下了灿烂的篇章，为近代苏州建筑的传承、弘扬、创新和融合提供了厚植浓厚的人文气息和历史底蕴。

《启迪设计集团作品集1953—2023》为我们徐徐展开启迪设计70年的发展画卷，历数一批批启迪设计人心血的设计作品，回溯时光，见证启迪设计在苏州镌刻下熠熠生辉的城建印记，描绘出波澜壮阔的时代画卷。

启迪设计及前身苏州市建筑设计研究院，作为中华人民共和国成立之初就设立的设计机构，70年来在不同时期的苏州城建中精耕细作，与苏州城市共生共荣、相辅相成，堪称“一部建院史、半座姑苏城”。从改革开放前的苏州早期十大建筑，到覆盖工业建筑、教育建筑、住宅建筑、剧院、商场、车站等多领域的设计作品，都是市民家喻户晓、耳熟能详的城市记忆，这些精品不仅代表着启迪设计这家“市院”的设计水平，更代表着苏州的城市建设水准，成功塑造出具有历史传承风骨和地域文化符号的苏州城市形象。

改革开放以来，启迪设计伴随着城市建设发展的历程创作出一大批优秀项目作品并获奖，许多苏州的地标建筑，如苏州中心、太湖文化论坛国际会议中心、东山国宾馆、苏州纳米城、苏州橙天360剧场、冯梦龙村、昆山杜克大学、太湖新城地下空间等均由启迪设计主持或参与完成。今天，苏州博物馆、苏州中银大厦、苏州国际博览中心、苏州科技馆与工业展览馆，已经成为见证苏州勇于开放并与国际接轨融合的标志性建筑。人民路综合整治提升工程、潘祖荫故居改造、书香世家·平江府、苏州会议中心改造提升、斜塘老街四期、树山村改造提升，则为我国的城市更新提供了难得的成功经验。

本书是启迪设计在苏州作品的集中展示，清晰而真实地反映了苏州城建的巨大变化，展现了70年来启迪设计从一间设计室到全国一流城乡建设科技集团的华丽蜕变，同时也是苏州从江南古城迈向现代化都市的重要见证。

中国工程院院士
东南大学教授

前言

2023年，是启迪设计集团股份有限公司（以下简称“启迪设计”）成立70周年。启迪设计的前身——苏州市建设局设计室是苏州第一家建筑设计院，从诞生起便肩负着建设苏州城市、传承苏州文化的重要使命。生与斯、长于斯，启迪设计与时代同脉搏、同命运，与苏州共奋进、共生长，书写出苏州现代化城市建设的壮丽史诗，沉淀为苏州城市发展的新历史。在启迪设计成立70周年之际，谨以此书献给一个个奋斗不止、自强不息的启迪设计人，献给苏州这座历久弥新、风华正茂的城市。

一部建院史，半座姑苏城。过去的70年，是苏州2500年历史画卷中浓墨重彩的一笔，苏州紧紧抓住每一次重大战略机遇，在发展中前进，在前进中突破，实现了“农转工”“内转外”“量转质”的华丽转变，创造出众多令人瞩目的“苏州奇迹”。在这些“奇迹”背后，矗立起的是一个个启迪设计的项目作品，汇聚着的是一批批启迪设计人的智慧与心血。

70年栉风沐雨，朝乾夕惕，启迪设计在中国现代化发展的丰沃土壤中生根发芽，在苏州城市化建设的春风化雨中枝繁叶茂，我们牢牢根植苏州城市文脉，矢志吴文化传承，尽展姑苏城风采，厚植千万苏州市民福祉，用传承创新之理念、唯精唯一的匠心，融江南水乡之情趣、园林之精巧、建筑之精湛，描绘出千年古城的历史积淀和现代风韵。

城市的发展离不开得天独厚的外在条件，更离不开砥砺奋进的内在精神。苏州自古以来就是江南富庶之地、园林之城，有历史之蕴，有人文之涵，有山水之韵，有建筑之美。70年来，一代代新老苏州人抢抓机遇、克难奋进、开拓创新，经济社会、城市建设、人民生活发展取得了令人瞩目的巨大成就，谱写了更加璀璨绚烂的现代篇章。让传承、创新、融合成为苏州转型发展、跨越腾飞的内在基因，也成为启迪设计的使命与担当。

70年后的今天，我们站在历史的交汇点深情回望，对每一段发展历程都无比珍视，对每一个项目创作都满怀敬意。无论在哪个发展阶段，启迪设计都始终牢记初心与使命，勇于追求卓越与非凡，不断为中国城市建设提供苏州方案、打造苏州范例、贡献苏州力量。

中华人民共和国成立后，苏州开启经济建设的探索实践，启迪设计作为苏州第一家建筑设计机构，在物质短缺、经济落后的条件下，义无反顾、全心全意投身城市建设，创作了苏州轻工机械厂、苏州钢铁厂、美术瓷厂、东方红造纸厂、苏纶纺织厂、东吴酒厂、跃进影剧场、新华书店等工业及文化领域的重要建筑。这些建筑凝聚着早期启迪设计人的智慧与汗水，不但在苏州各领域实现了诸多“第一”的创举，也催生了如苏州早期十大建筑等一大批优秀作品，为苏州市民带来了一座座印刻在记忆中的建筑，努力推动苏州实现经济发展和城市建设的历史跨越。

改革开放以来，苏州创新发展“苏南模式”，城市建设飞速发展。启迪设计乘着改革春风，广泛吸收先进设计经验，积极投入苏州现代化城市建设发展中，在苏州各区、县均有大量优秀作品：改扩建后升格为客货一等站、获评苏州最优秀建筑的苏州火车站，以“建筑风格园林化、内部功能现代化”为特点的苏州图书馆，承载着苏州市民健康愿景的苏州体育中心，为苏州妇幼保健做出重大贡献的苏州母子医疗保健中心，苏州规模最大的汽车客运站汽车南站，获国家建设部优秀设计银质奖的彩香住宅小区，熔古铸今的标志性公共建筑——苏州市规划展示馆，苏州市民和国内外游客耳熟能详的第一百货、人民商场、泰华商城，苏州大学的标志性建筑——理工实验大楼，市政府大楼以及中国银行、中信银行等金融机构。

同时，一批镌刻时代记忆、声名远播的优秀建筑，如与国际建筑大师贝聿铭合作的苏州博物馆，已成为苏州著名的文化元素和城市名片。还有获得中国旅游业标志性饭店金奖的竹辉饭店、获得国家优秀设计金奖的苏州刺绣研究所、苏州城区第一高楼雅都大酒店、中国第一个专业性质的丝绸博物馆、系统展现苏州革命建设伟大历程的革命博物馆、为苏州文物事业发展做出巨大贡献的苏州文物商店等。这些建筑是记录“苏南模式”快速崛起光辉历史的活化石，也成为苏州在世纪之交高速发展的跨时代记忆。

90年代苏州紧紧抓住全国改革开放重心从珠三角移向长三角地区的重大历史机遇，尤其是苏州工业园区设立后，在中新合作发展的窗口机遇下，集团勇担城建拓新先锋，率先投身园区规划建设，完成了诸多“第一”：打下园区第一根桩，完成园区首个项目——腾飞新苏工业坊，园区第一栋高层——馨都广场，园区第一栋外资厂房——纳贝斯克，拥有8000m^{2}无柱宴会厅、逾50000m^{2}连续无柱多功能展厅的园区最大单体建筑——苏州国际博览中心，园区第一个住宅小区——新城花园，园区第一个轨道交通地下空间——星海生活广场，园区第一个绿色三星运营建筑——启迪设计办公楼，园区第一个绿色三星运营建筑+LEED铂金级认证——宝时得中国总部，园区第一条传统风貌商业街区——斜塘老街，园区首家“医养融合”公办护理院——久龄公寓，园区湖东第一栋建筑——东湖大郡，园区独墅湖科教创新区第一栋建筑——苏州研究生城综合楼，独墅湖地区最高建筑——中新大厦。

苏州作为历史文化名城，是一座传统与现代交织的城市，长期以来高度重视古城保护，城市更新一直走在全国最前列。启迪设计立足苏州文脉特征，依托全产业链集成服务优势，传承发扬苏式工匠精神，积极探索城市更新模式、不断激发城市活力。在20世纪90年代完成了全国首批旧城改造试点工程项目——桐芳巷小区，桐芳巷小区曾获全国优秀工程勘察设计行业奖综合金牌奖，以桐芳巷为代表的“新建街区，风貌延续”成为后来苏州古城更新的主要模式。此后启迪设计对苏州观前街的整治改造，进一步奠定了观前街“姑苏第一街”的地位。南浩街作为苏州城区传承800余年的民俗活动——“轧神仙”的场所，经过启迪设计改造后也成功保留既有建筑风貌和人文景观，并获得2000年国家建设部表彰。

启迪设计长期在街区综合提升、既有建筑更新、传统建筑活化、老旧小区改造等多个领域深入研究，于长期实践中积累了丰富的成功经验。苏州入选全国首批城市更新试点城市后，启迪设计紧抓机遇，加快探索城市更新新方案、新路径，大力推广城市更新成功经验，推动苏州高质量发展，并持续为全国城市更新工作贡献更多的苏州经验。

苏州是首批国家生态园林城市，与下辖四市、五区均为全国生态文明建设试点地区，在长三角区域一体化发展中扮演着重要角色。得益于此，在助力实现碳达峰目标、推动绿色发展的道路上，启迪设计务实笃行、行稳致远。

在生态景观建设方面，无论是坐拥旖旎湖光的西京湾、丽波湾、阳澄湖、杵山，连接苏州工业园区湖东湖西的金鸡湖大道、现代大道，还是金鸡湖游人如织的月光码头和斜塘老街、风景怡人的活力岛、鸷山桃花源，启迪设计都以匠心营造生态景观，为苏州这座城市塑造出宜居宜业的优美形象。

在生态修复与海绵城市建设方面，有长三角生态绿色一体化示范区内的首个滨水生态工程——元荡先导段生态修复项目；将海绵技术运用在景观绿化中，实现糅合景观的生态修复，并成为苏州旅游发展新支点的苏州乐园森林世界；将多元化的海绵元素深度融入景观设计的三江口公园。

在绿色低碳建筑建设方面，启迪设计完成了获得美国LEED-NC铂金级、中国国家绿色建筑三星级设计标识双认证的宝时得中国总部；融合绿色健康智慧的国家级高品质示范标杆建筑、江苏省高品质绿色建筑示范项目、三星级绿色建筑——启迪设计大厦；获得江苏省绿色建筑创新一等奖的常熟世联书院；获得年度十大绿色建筑项目的太仓裕沁庭；获得近零能耗建筑标识证书、江苏省低碳建筑科技示范工程的苏高新绿色低碳示范基地；全国首个获得绿色建筑三星级标识认证的独立式地下空间工程——苏州太湖新城核心区地下空间。

时至今日，当我们从苏州古城最高点北寺塔眺望，视线所及，穿越古今，从一片粉墙黛瓦的古城，到高楼林立的新城，从烟波浩渺的太湖之滨，到波澜壮阔的长江之畔，处处都散发着苏州古今交织、中外合璧的独特魅力，昭示着启迪设计人在不同年代、不同地点为这座城市贡献的点滴力量，汇聚成苏州城市的精神气质，谱写出启迪设计跨越两世纪、横亘七十载的城建史诗。

董事长致辞

苏州自古便是钟灵毓秀之地，有历史之蕴，有人文之涵，有山水之韵，有建筑之美。启迪设计集团股份有限公司作为苏州市第一家建筑设计院，便诞生在这片丰沃的土壤之中，栉风沐雨，朝乾夕惕，一步步成长为行业领先、全国一流的城乡建设科技集团。

翻开集团70年历史，我们身后，是三千年的吴文化，是两千年的姑苏城，是一千万的苏州市民，我们与祖国共生、共长、共荣。70年来，前辈们筚路蓝缕，披荆斩棘，为城市建设做出了巨大贡献，苏州今日的成就是经济发展的真实写照，也凝聚着数代苏州勘察设计工作者的毕生心血，书写了启迪设计集团在城市建设中的壮丽史诗。

1953年，在中华人民共和国成立之初，百废待兴，集团前身——苏州市建设局设计室应运而生，从诞生起便背负着建设苏州城市、传承江南文化的重要使命。改革开放以来，乘着发展春风，集团投身于苏州城市建设，给苏州市民带来了一座座记忆中的建筑，为这座古城的发展留下了浓墨重彩的一笔。90年代苏州工业园区设立后，在中新合作发展的窗口机遇下，在苏州经济快速腾飞的背景下，集团勇担建设先锋，投身园区规划建设，在园区打下第一桩，打造出国际合作的成功范例。

进入新世纪后，启迪设计大力创新、率先转型，顺利改制，以民营科技企业的崭新姿态开启新世纪发展的新阶段，全力参与支持苏州各区市及全国其他地区的经济和城市建设。党的十八大以来，我国全面建设小康社会进入决定性阶段，国家发展进入新时期，启迪设计紧抓时代机遇，稳步实现股改、上市的目标，在资本市场的推动下，开启快速扩张与发展的崭新征程。

时至今日，启迪设计业务已由单一的建筑设计，向覆盖人居环境全过程的投融资、咨询、设计、建造、运维等多元化全产业链集成服务发展，形成了以“全过程咨询+工程建设管理+双碳新能源+城市更新+数字科技”五大板块为支撑的业务格局，全方位紧跟和服务国家战略，不断向“全国一流城乡建设科技集团”的目标迈进。集团拥有包括建筑行业甲级在内的多项设计资质和施工资质，是国家高新技术企业、中国十大民营工程设计企业、国家首批装配式建筑示范产业基地、国家全过程工程咨询试点企业。

集团旗下已拥有深圳嘉力达、深圳毕路德、中正检测等多家控股（全资）子公司及分公司，在巩固江苏省内本土市场的同时，加快全国化布局，大力开拓长三角、大湾区、京津冀、成渝陕、湘鄂赣等五大区域市场，建成了华东、华北、华南、西南、华中等多区域的全国性服务网络，业务版图覆盖全国，项目作品遍布各地。

集团汇聚千余名优秀人才，涵盖享受国务院政府特殊津贴专家、江苏省设计大师、全国优秀勘察设计行业优秀企业家、中国建筑学会青年建筑师奖获得者、长三角建筑学会联盟青年建筑师奖获得者、省突出贡献的中青年专家、省“333工程”人才、省优秀工程勘察设计师等高层次人才。

集团拥有省级工程技术研究中心——江苏省（赛德）绿色工程技术研究中心、江苏省智慧园区系统集成工程研究中心、建设领域双碳技术全过程创新应用中心。与清华大学、东南大学、同济大学等多所重点院校建立校企合作，并成为省博士后创新实践基地、国家博士后科研工作站。相继成立数字科技事业部、双碳新能源事业部、建筑结构技术创新研究院、智能化建筑应用研究中心、苏州思萃城市更新产业技术研究所等多个专业技术中心，全方位开展科技研发。

70年的流金岁月，70年的奋发图强，70年的跨越腾飞。在党的二十大精神感召下，集团牢记传承、保护和发扬中华民族建筑文化的使命，锐意进取，主动识变、应变、求变，以勇于融合、创造人居环境美好未来为己任，坚持科技创新引领发展，将启迪设计发展成为勘察设计行业领先、特色领域技术优势明显、综合实力雄厚的全国一流城乡建设科技集团。以实际行动谱写新时代更加绚丽的华章，奋力推动城市建设、行业进步高质量发展，为人民创造高品质生活贡献智慧和力量，为全面建设社会主义现代化国家、全面推进中华民族伟大复兴不懈努力、奋斗。

启迪设计集团　党委书记
董事长
首席总工程师

总裁致辞

苏州作为吴文化的发祥地，兼具千年古城之深厚历史文化底蕴，又融入现代化都市的规划建设理念，是一座真正意义上熔古铸今的城市。我们生与斯、长于斯，70年的夙兴夜寐，跨世纪的传承创新，启迪设计用一个个项目作品，在城市中烙下发展的印记；用一张张照片图纸，在历史中留下独特的色彩。

苏州以其“水陆并行、河街相邻”的城市格局、“小桥流水、粉墙黛瓦”的独特风貌、“烟波浩渺、江南胜境”的湖光山色，感染着古今的文人墨客，厚重了苏州的地域文化。从苏式建筑的角度看，枕河的民居、重重的院落、精雅的园林，无一不是可供后世汲取的元素，苏州建筑的根由此而来，苏州城市的魂由此而生。

启迪设计深植苏州千年文化土壤，在苏州博物馆、苏州刺绣研究所接待馆、斜塘老街四期、潘祖荫故居改造、书香世家·平江府酒店、星海街9号厂房改造等项目中，我们或新建，或改造，提炼苏式传统文化元素与太湖秀丽山水神韵，将之融入项目建造，从建筑设计的角度，传承苏式建筑，延续苏州文脉。

苏州不仅文化旅游资源丰富，更是中新合作发展的成功案例，是长三角对接国际的窗口，我们参与的苏州中银大厦、昆山杜克大学、苏州太湖新城地下空间（中区）等项目，不仅覆盖城市建设的多个领域，更开展了形式多样、成绩斐然的国际合作，古今在此交融，中外在此汇聚。

经济的腾飞，为苏州带来了人口的爆发式增长，也带来了对文化产业的迫切需求，苏州图书馆、苏州工业园区北部文体中心、苏州橙天360剧场、苏州国际博览中心、苏州科技馆及工业展览馆、冯梦龙村全域旅游发展规划等项目，进一步为城乡增添文化艺术气息，让工业与艺术之美在千年古城交融，让美丽田园乡村建设绽放苏州光彩。

我们的脚步从未停下，随着集团规模与业务版图的不断拓展，走出苏州，迈向全国，在更广的区域市场布局发展，在更多的建设领域开展实践，以创意引领设计，以创新推动技术，以覆盖人居环境全过程的多元化全产业链集成服务，实现了在建设领域的跨越式发展。

苏州中南中心、恒力全球运营总部、苏州自贸商务中心、中新大厦、苏州港口发展大厦、文旅万和广场等项目不断刷新新高度、新规模。在商业建筑领域，苏州中心广场、丰隆城市中心、南通圆融广场、龙湖狮山天街、苏悦广场、常熟永旺梦乐城、月光码头等一批项目取得了良好反响。第九届江苏省园艺博览会工程B馆、苏州妇女儿童活动中心、太湖文化论坛国际会议中心、苏州慈济园区、苏州评弹公园、苏州阳澄国际电竞馆等文体建筑项目已成为苏州城市多元文化的名片与形象。

苏州金普顿竹辉酒店、太湖万丽万豪酒店、东山宾馆、裸心泊酒店等项目，或依古城风貌，或借太湖风光，缔造出一个个别具特色的文旅酒店。苏州大学理工实验楼、中国科学技术大学苏州研究院仁爱路校区、中国中医科学院大学、南通大学蔷园校区、西交利物浦大学行政信息楼、常熟世联书院、苏州木渎高级中学、苏州实验中学科技城校、吴郡幼儿园等教育建筑，横跨各教育阶段，进一步夯实苏州教育基础，巩固经济发展成果。

无锡医疗健康产业园、国家区域中医诊疗中心、河南省中医院、苏州大学附属第二医院应急急救与危重症救治中心、苏州市立医院康复医疗中心、苏州工业园区久龄公寓等医疗康养项目，覆盖医养建筑全领域、全过程，成为集团独具优势与特色的业务板块。微软（中国）苏州科技园区二期、枫桥工业园、金唯智基因组研究和基因技术应用实验楼、骊住科技新工厂、大兆瓦风机新园区项目、苏州纳米城、宝时得中国总部一期等科技研发建筑，植入更多科技元素，重新定义工业建筑。

人民路综合整治提升工程、苏州市政府大楼、苏州会议中心改造提升、明城墙沿线旧城出新项目、树山村改造提升、消泾村二亩塘特色田园乡村等既有建筑改造和乡村振兴项目，凸显苏州地域文化特色，成为国内借鉴交流的典范与模板。

我们追忆往昔、耕耘当下、翘首未来，将70年青春的活力、智慧的光辉、拼搏的激情，辑录成书，这是几代人践行使命的光荣见证，是几十年接力发展的伟大史诗，也是照亮未来的荣耀篇章。我们以梦为马，谱写了70年的辉煌历史，也即将踏上一个崭新的奋斗征程。

启迪设计集团　总裁/首席总建筑师

江苏省设计大师

目录

02

03

01

代表作品

REPRESENTATIVE PROJECTS

启迪设计大厦
TUS-DESIGN BUILDING
星海街 9 号厂房改造
RECONSTRUCTION OF NO. 9 PLANT IN XINGHAI STREET
苏州博物馆
SUZHOU MUSEUM
苏州中银大厦
SUZHOU BANK OF CHINA
江苏银行苏州分行
BANK OF JIANGSU SUZHOU BRANCH
苏州国际博览中心
SUZHOU INTERNATIONAL EXPO CENTER
苏州科技馆与工业展览馆
SUZHOU SCIENCE HALL AND INDUSTRIAL EXHIBITION HALL
苏州橙天 360 剧场
SUZHOU OSGH 360 THEATER
江苏硅谷展示馆
JIANGSU SILICON VALLEY EXHIBITION HALL
苏州工业园区北部文体中心
SIP NORTHERN CULTURE AND SPORTS CENTER
潘祖荫故居改造
RECONSTRUCTION OF PAN ZUYIN'S FORMER RESIDENCE

书香世家 · 平江府酒店
SCHOLARS HOTEL SUZHOU PINGJIANGFU
苏州太美 · 逸郡酒店
G-LUXE BY GLORIA SUZHOU TAIMEI
斜塘老街四期
XIETANG OLD STREET PHASE Ⅳ
无锡尚德太阳能电力研发办公楼
WUXI SUNTECH R&D OFFICE BUILDING
黄山小罐茶运营总部
HUANGSHAN XIAO GUAN CHA OPERATION HEADQUARTERS
冯梦龙村全域旅游发展规划
FENGMENGLONG VILLAGE OVERALL TOURISM DEVELOPMENT PLAN
昆山杜克大学二期
DUKE KUNSHAN UNIVERSITY PHASE Ⅱ
中国中医科学院西苑医院苏州医院
SUZHOU HOSPITAL OF XIYUAN HOSPITAL OF CHINA ACADEMY OF CHINESE MEDICAL SCIENCES
无锡医疗健康产业园
WUXI MEDICAL HEALTH INDUSTRIAL PARK
苏州太湖新城地下空间（中区）
SUZHOU TAIHU NEW TOWN UNDERGROUND SPACE (MIDDLE AREA)

启迪设计大厦

TUS-DESIGN BUILDING

项目类型： 总部办公　　**项目地点：** 江苏苏州
用地面积： 15678m²　　**建筑面积：** 78477m²
设计时间： 2019年　　**竣工时间：** 2023年

绿色建筑三星认证
健康建筑三星级认证
LEED金级国际绿色标识认证
能效测评三星级认证
绿色建筑创新奖
中国建设工程鲁班奖
第八届江苏省勘察设计行业信息模型（BIM）应用大赛一等奖

启迪设计大厦作为启迪设计集团新总部大楼，是中国（江苏）自贸试验区苏州片区正式挂牌后奠基动工的第一栋总部大楼。2020年本项目获批江苏省高品质绿色示范项目，因其集建设方、设计方、运维使用方“三位一体”的独特性，成为全过程落实传统文化精神与绿色低碳理念的实践案例。

As the new headquarters building of the Tus-Design Group, the Tus Design Building is the first headquarters building to start construction after the official listing of the Suzhou area of the China (Jiangsu) Pilot Free Trade Zone. In 2020, this project was approved as a high-quality green demonstration project in Jiangsu Province. Due to its unique integration of construction, design, and operation, it has become a practical case for implementing traditional cultural spirit and green and low-carbon concepts throughout the entire process.

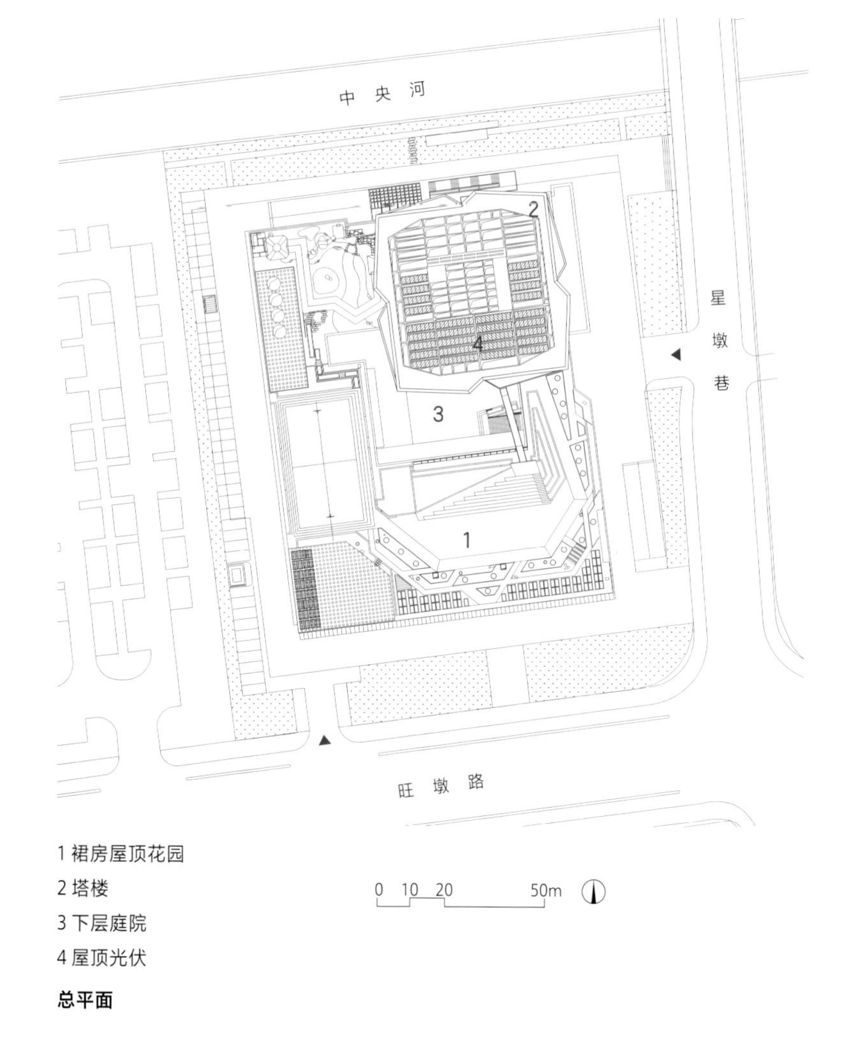

1 裙房屋顶花园
2 塔楼
3 下层庭院
4 屋顶光伏

0 10 20 50m

总平面

区位

启迪设计

启迪设计大厦设计中，将高层塔楼在垂直方向上分成七个单元体，每四层形成一个单元体，每个单元体设置一个三层通高的共享空间和一层四周收进的阳台区域，三层通高的共享空间构成了高层中的“垂直院落”，而四周收进的阳台区域宛如“空中游廊”，“垂直院落”与“空中游廊”叠加而成的多层空间格局，呼应了苏州院落与园林中环廊的空间。

In the design of the building, the high-rise tower is vertically divided into seven units, each with four floors, and each has a three-story shared space and a balcony area. The shared spaces form a "vertical courtyard" in the high-rise, while the surrounding balcony area is like a "sky corridor". "Vertical courtyard" and "sky corridor" are stacked in a multi-level spatial pattern that echoes the space of the ring corridor in Suzhou courtyards and gardens.

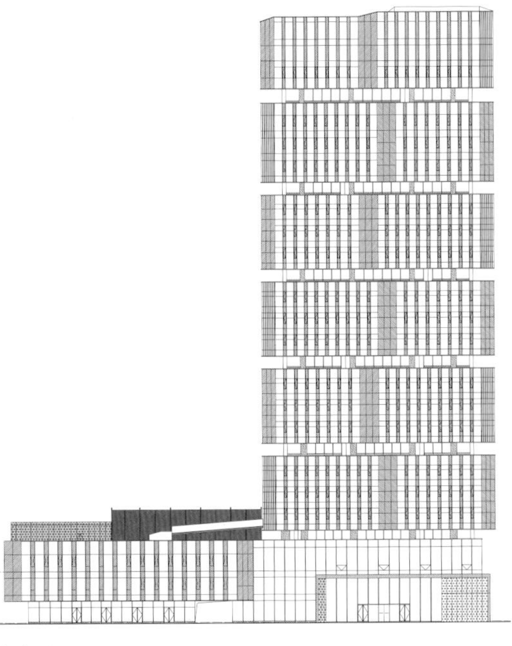

东立面

应对可持续发展的理念和绿色建筑的发展趋势，新大楼运用了大量先进的绿色技术，在持续降低能耗的同时更多关注使用者的健康与舒适，智能化的运维系统使绿色建筑从粗放型向精细化转变，达到高效绿色的可持续应用。

The design of the building responds to the concept of sustainable development and the trend towards green buildings, the new building utilizes a large number of advanced green technologies that continue to reduce energy consumption while paying more attention to the health and comfort of the users. Intelligent operation and maintenance systems enable green buildings to transform from crude to refined to achieve efficient green and sustainable applications.

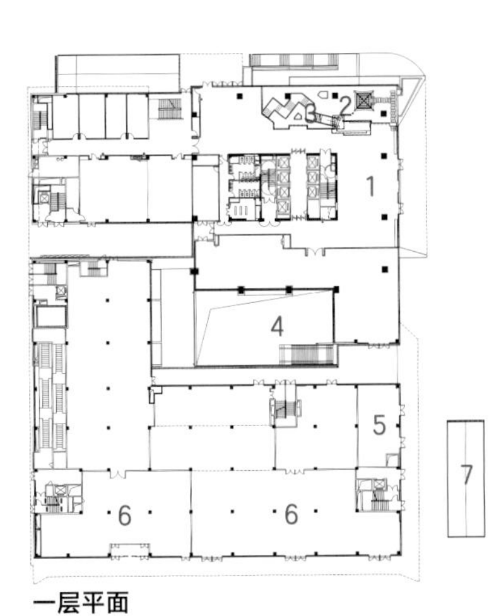

1 塔楼门厅
2 景观水池
3 旋转楼梯
4 下沉庭院
5 超市
6 银行营业厅
7 汽车坡道

一层平面

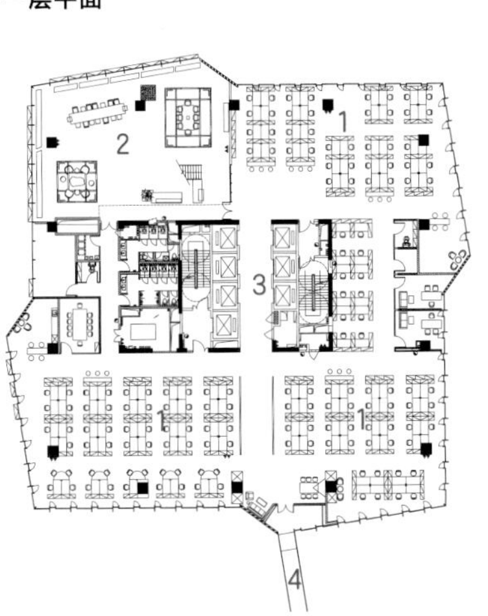

1 办公区域
2 共享空间
3 电梯厅
4 天桥

五层平面

星海街9号厂房改造

RECONSTRUCTION OF NO. 9 PLANT IN XINGHAI STREET

项目类型： 既有建筑改造/办公　　**项目地点：** 江苏苏州
用地面积： 18547m²　　**建筑面积：** 11552m²
设计时间： 2009年　　**竣工时间：** 2010年

全国三星级绿色建筑运行标识项目
2013年度全国绿色建筑创新奖二等奖
2011年度华夏建设科学技术奖
第五届全国民营工程设计企业优秀工程设计-园林和景观工程类华彩铜奖
2015年度全国优秀工程勘察设计奖绿色建筑二等奖
2012年度江苏省第十五届优秀工程设计一等奖
2015年度江苏省优秀工程勘察设计行业奖绿色建筑专业一等奖

星海街9号项目改造前原有建筑为一个面积为6800m²的单层工业厂房。设计保留了原有建筑轮廓和全部结构，利用原有厂房较高层高的特点，将整个空间分为两层。建筑平面中间设置两个中庭，将原有的屋面固定采光孔改造为可开启的通风采光孔，南立面和西立面增加生态遮阳走廊等一系列措施，有效地解决了自然通风、自然采光和生态遮阳。

The original building before the renovation of the No. 9 Xinghai Street project was a single-story industrial building with an area of 6,800 square meters. The design retains the original building outline and all the structure, and utilizes the higher floor height of the original factory building to divide the entire space into two floors. A series of measures such as setting up two atriums in the middle of the building plan, transforming the original fixed roof lighting holes into openable ventilation and lighting holes, and adding ecological sun-shading corridors on the south and west facades effectively solve the problems of natural ventilation, natural lighting and ecological sun-shading.

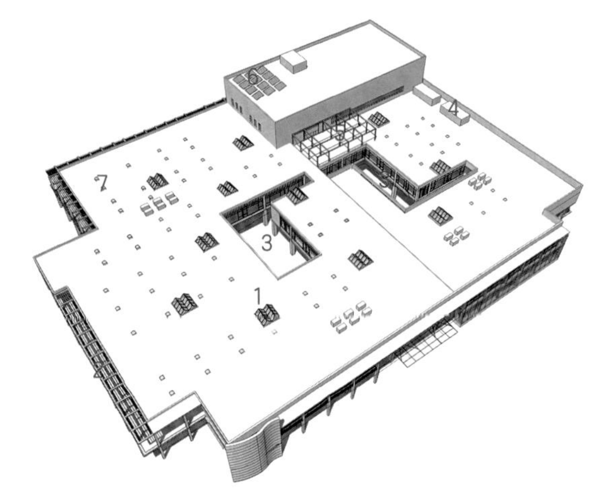

1 电动开启天窗
2 双层玻璃幕墙系统
3 内庭院
4 厨房油烟净化处理系统
5 屋顶绿化实践区
6 太阳能热水系统
7 日光照明系统

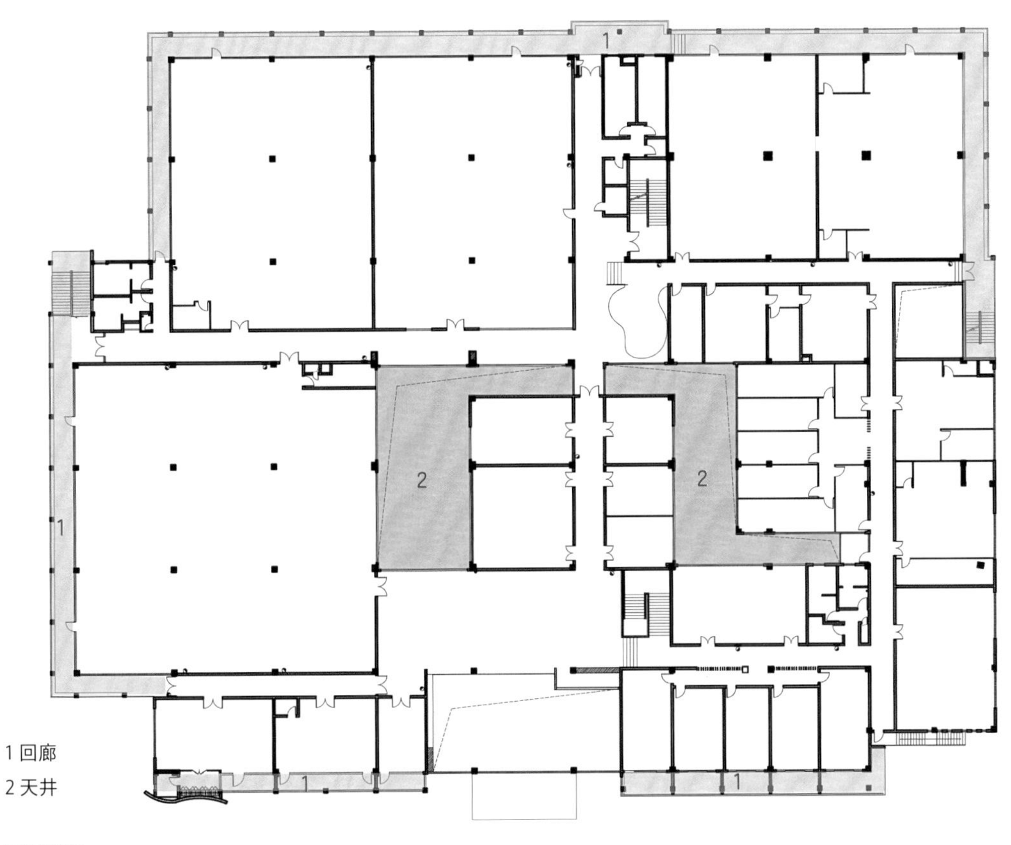

二层平面

与自然的融合共生，建筑的改造为植物生长提供空间和平台。
Integration and symbiosis with nature, the reconstruction of buildings provides space and platform for plant growth.

通过对既有生产厂房的更新改造，运用绿色科技产品手段，融入了高效健康的人性化使用空间，延长了建筑寿命，最大限度地节约资源，创造了绿色生态设计的典型范例。这座再生的工业厂房，绿色、节能、生态、环保概念贯穿其中，实现了生态技术及本土文化价值的双重传导。它既是启迪设计绿色创作理念与生态技术集成的展示平台，也是企业设计文化的标志与象征。

Through the renewal and renovation of existing production plants, the use of green technology products means to incorporate efficient and healthy humanized use of space, extend the life of the building, maximize resource conservation, and create a typical example of green eco-design. The concepts of green, energy-saving, ecology and environmental protection run through this regenerated industrial building, realizing the dual transmission of ecological technology and local cultural values. It is not only a platform for displaying Tus-Design's green creative concepts and eco-technology integration, but also a sign and symbol of the enterprise's design culture.

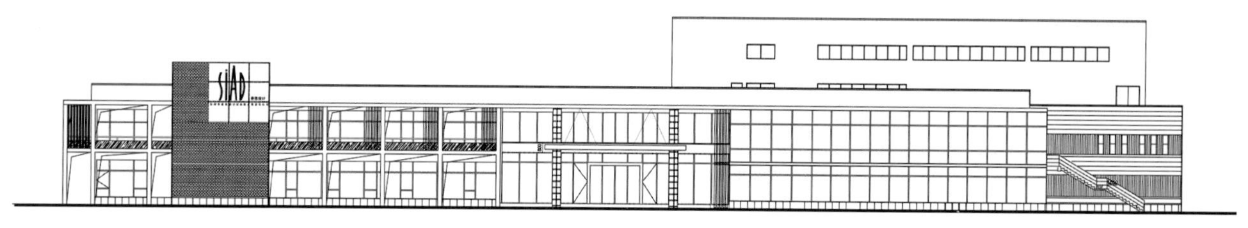

东立面

西立面

改造后的办公楼被重重绿意包裹，开放庭院贯穿其中，藤架长廊悬于外墙，高大林木环立四周。在封闭的厂房建筑中，架构起由内而外的三重自然空间，江南的传统气质自然彰显。这些无处不在的绿色空间，提供了人与自然亲密接触的可能。在这里，可以放松心情愉悦工作，无限创意随时迸发。

The renovated office building is enveloped in a heavy greenery, with an open courtyard running through. The pergola corridor overhangs the exterior wall and tall trees surround the office building. In the enclosed factory building, a triple natural space from inside to outside is constructed, which naturally highlights the traditional temperament of Jiangnan. These ubiquitous green spaces provide the possibility of intimate contact between humans and nature. In this place, employees can relax and work happily, bursting unlimited creativity at any time.

苏州博物馆

SUZHOU MUSEUM

项目类型：文化建筑　　**项目地点**：江苏苏州
用地面积：11000m²　　**建筑面积**：17000m²
设计时间：2003年　　**竣工时间**：2006年
合作单位：贝聿铭及贝氏建筑事务所

2019年度新中国成立七十周年优秀勘察设计优秀项目
2008年度全国优秀工程勘察设计奖金奖
2008年度全国优秀工程勘察设计行业奖一等奖
2009年度中国建筑学会建筑创作大奖
2012年度全国百项建筑智能化经典项目一等奖
2008年度江苏省第十三届优秀工程设计一等奖
2007年度江苏省城乡建设系统优秀勘察设计一等奖
2009年度江苏省优秀勘察设行业奖暖通二等奖
2009年度江苏省优秀勘察设行业奖结构二等奖
2009年度江苏省优秀勘察设行业奖智能化二等奖

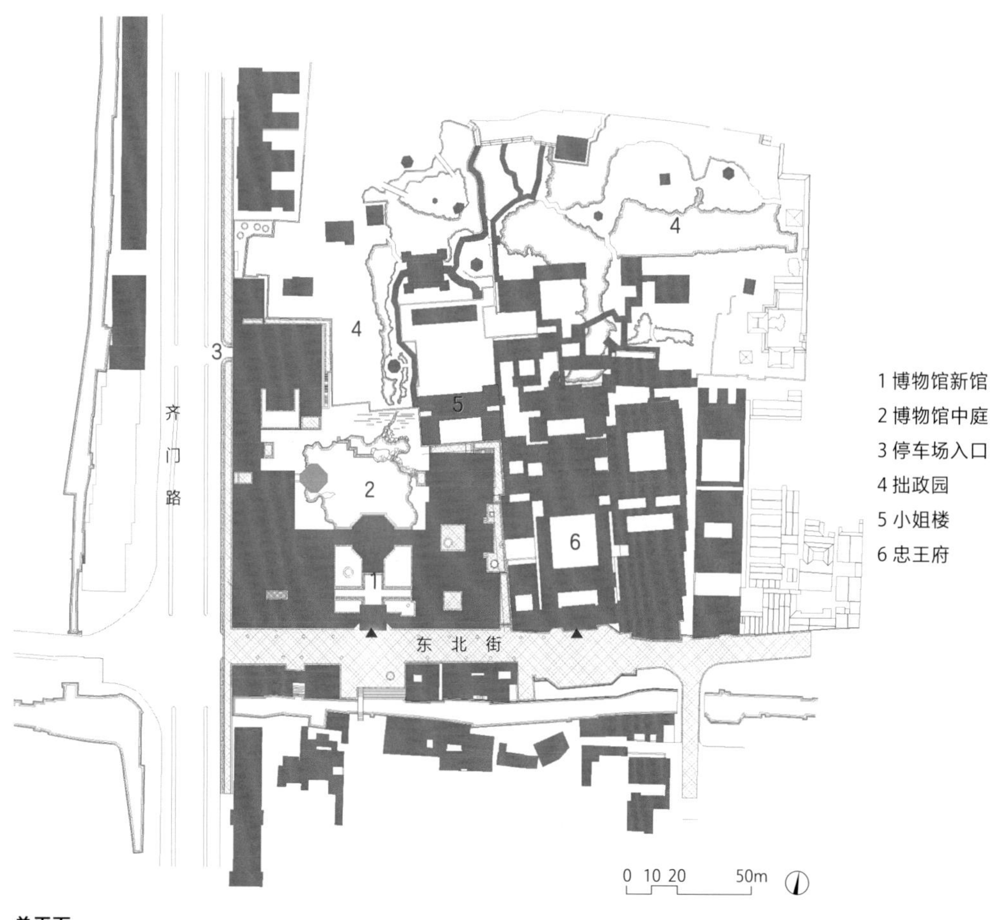

总平面

苏州博物馆位于苏州古城的历史街区中心，毗邻世界文化遗产拙政园、狮子林和全国重点文物保护单位太平天国忠王府。根据城市规划研究，设计任务的核心是把该地块打造成一个艺术文化区，将拙政园和狮子林连接在一起，为游客提供一条更为紧凑和终点明确的旅游线路。而“修旧如旧”的忠王府古建筑作为苏州博物馆新馆的一个组成部分，与新馆建筑珠联璧合，从而使苏州博物馆新馆成为一座集现代化馆舍建筑、古建筑与创新山水园林三位一体的综合性博物馆。

在整体布局上，项目巧妙借助水面，与紧邻的拙政园、忠王府相连，成为原有建筑风格的延伸。建筑群坐北朝南，分成三大块：中部为入口、中央大厅和主庭院；西部为博物馆主展区；东部为次展区和行政办公区。博物馆包括7500m^{2}展品陈列室、一个150座礼堂、办公区、文物保护工作室、研究图书馆及库房。

The Suzhou Museum is located in the center of the historical district of the Suzhou, adjacent to the Humble Administrator's Garden, a world cultural heritage, and the Taiping Heavenly Kingdom Zhong King Mansion, a national key cultural relics protection unit. According to urban planning research, the core of the design task is to create the plot into an artistic and cultural area, connecting the Humble Administrator's Garden and Lion Forest Garden, providing tourists with a more compact and obvious tourist route. As a component of the new Suzhou Museum, the ancient architecture of the Zhong King Mansion, which has been restored as old, is perfectly matched with the new building. Therefore, the new Suzhou Museum becomes a comprehensive museum integrating modern museum buildings, ancient buildings and innovative landscape gardens.

In the overall layout, the new Suzhou Museum skillfully utilizes the water surface to perfectly match the adjacent Humble Administrator's Garden and Zhong King Mansion, becoming an extension of its architectural style. The building complex faces south and is divided into three large sections: the central part is the entrance, central hall, and main courtyard; the west is the main exhibition area of the museum; the east consists of a secondary exhibition area and an administrative office area. The museum includes 7500m^2 exhibition rooms, a 150 seats auditorium, office area, cultural relics protection studios, research libraries, and warehouses.

博物馆在体量上遵循的原则是“不高、不大、不突出”，设计理念上追求“中而新、苏而新”。设计中大量新技术、新材料的运用，使这组新建筑既具有苏州园林传统特色，又处处散发着时代气息。

The principle followed by The Suzhou Museum in terms of volume is 'not high, not large, not prominent', in terms of design idea, designer pursues "the combination of Chinese characteristics, Suzhou flavor, and modernization". The application of a large number of new technologies and materials in the design not only embodies the traditional characteristics of Suzhou gardens, but also exudes the spirt of the times everywhere.

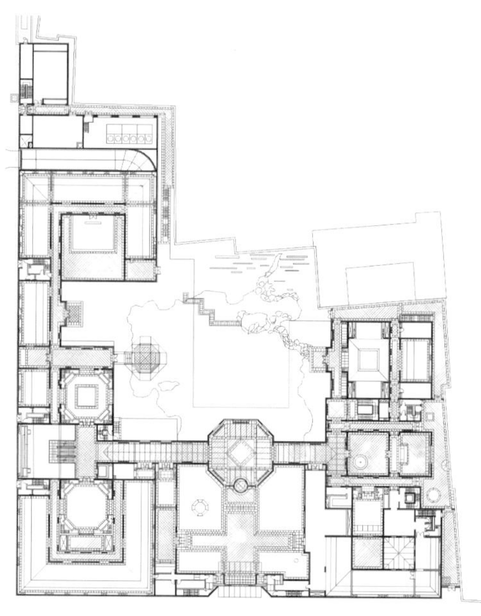

一层平面

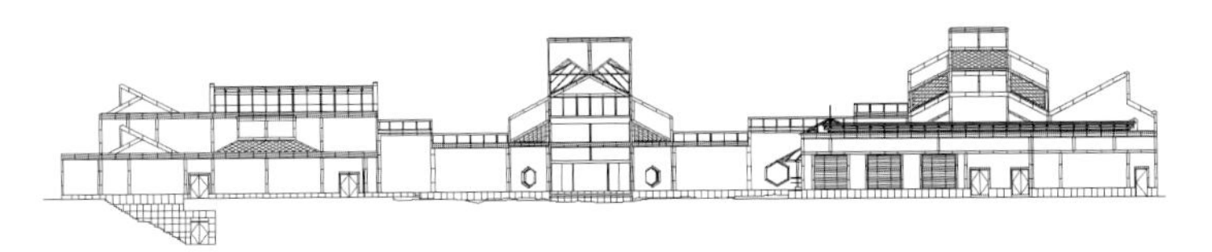

北立面

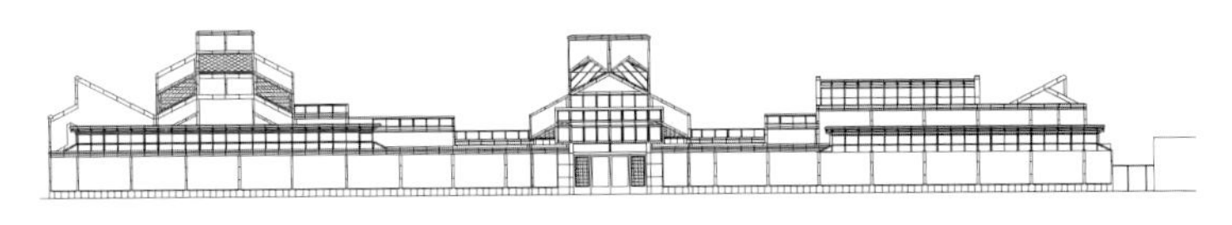

南立面

苏州中银大厦

SUZHOU BANK OF CHINA

项目类型： 金融办公　　**项目地点：** 江苏苏州
用地面积： 25096m²　　**建筑面积：** 99640m²
设计时间： 2013年　　**竣工时间：** 2015年
合作单位： 贝聿铭及贝氏建筑事务所

2017年度全国优秀工程勘察设计行业奖一等奖
第九届全国优秀建筑结构设计奖一等奖
2016年度江苏省第十七届优秀工程设计一等奖

中银大厦为中国银行股份有限公司苏州分行的总部大楼，坐落在苏州工业园区金融中心内。设计采用清晰强烈的几何体框架，营造充满日光的入口大厅，同时延续中国银行北京总行室内的石材贴面传统、贝氏丰富的方格天花等，展示中国银行的现代性、前瞻性、持续发展性。

This project is the headquarters building of Bank of China Limited Suzhou Branch, located in the financial center of Suzhou Industrial Park. The design utilizes a clear and strong geometric framework to create a daylight-filled entrance hall, while continuing the tradition of stone veneer and PEI's rich coffered ceilings in the interior of the BOC's Beijing Head Office, showcasing the BOC's modernity, forward-thinking, and sustainable development.

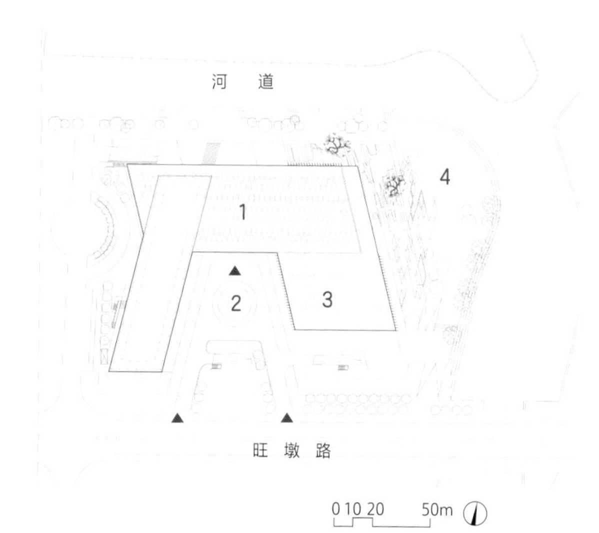

1 中银大厦　　3 景观水体
2 主入口广场　　4 景观广场

总平面

SIPAB

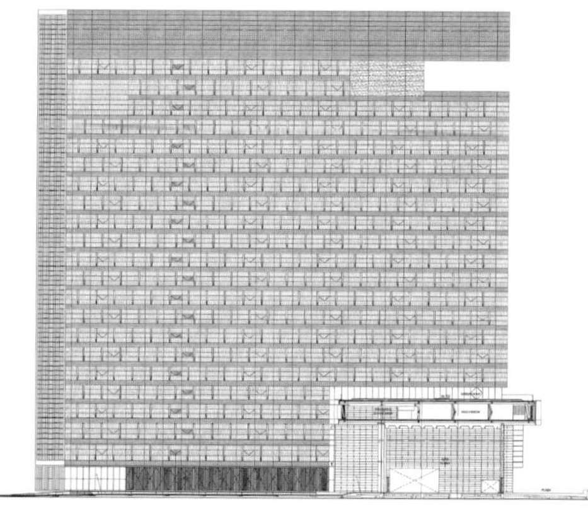

东立面

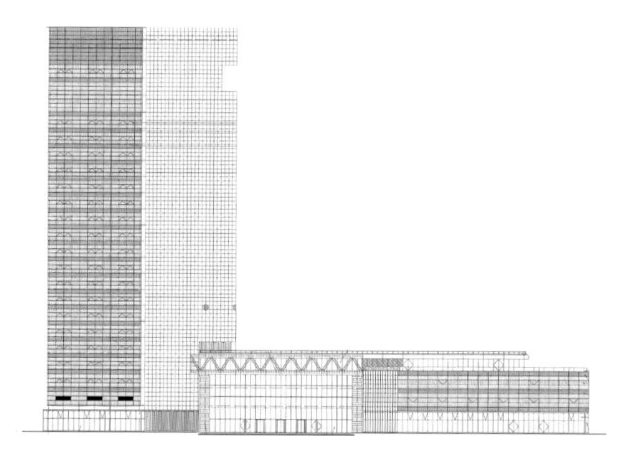

南立面

设计利用其基地中最显著的景观河道优势，沿河区域参考苏州古城传统的河道设计手法设置滨河走廊。立面造型是对苏式传统建筑的全新演绎，传统的灰和白的主色调被充分利用，城市公共滨水空间也被统一融入到景观设计中。

The design takes advantage of the most significant landscape river in its base, and sets up a riverfront corridor along the river area with reference to the traditional river design of the ancient city of Suzhou. The facade shape is a new interpretation of traditional Suzhou architecture, the traditional gray and white primary colors are fully utilized, and the urban public waterfront space is unified into the landscape design.

江苏银行苏州分行

BANK OF JIANGSU SUZHOU BRANCH

项目类型： 金融办公　　**项目地点：** 江苏苏州
用地面积： 5120m²　　**建筑面积：** 48588m²
设计时间： 2010年　　**竣工时间：** 2015年

2017年度全国优秀工程勘察设计行业奖三等奖
2016年江苏省城乡建设系统优秀勘察设计一等奖
2016年度江苏省第十七届优秀工程设计二等奖

项目位于苏州园区湖西CBD核心区，设计充分考虑与周边城市环境的协调与融合，借鉴江苏地区的建筑元素和文化特征，以符合当地特色的建筑形态和立面设计。设计受江苏银行行徽的启发，突破周边方盒子式的高层建筑群的传统造型，建筑形体通过两个"J"形体量旋转生成，两个体量互相环抱、不断攀升，象征着企业蒸蒸日上的积极形象。

Bank of Jiangsu Suzhou Branch is located in the CBD core area of Suzhou Industrial Park. The design gives full consideration to the coordination and integration with the surrounding urban environment, and draws on the architectural elements and cultural characteristics of the Jiangsu region in order to conform to the local characteristics of the building form and facade design. Inspired by the emblem of Bank of Jiangsu, the design breaks away from the traditional shape of the neighboring box-like high-rise complexes, and the architectural form of Bank of Jiangsu Suzhou Branch is generated by the rotation of two "J" volumes, which embrace each other and continuously climb, symbolizing the positive image of the thriving enterprise.

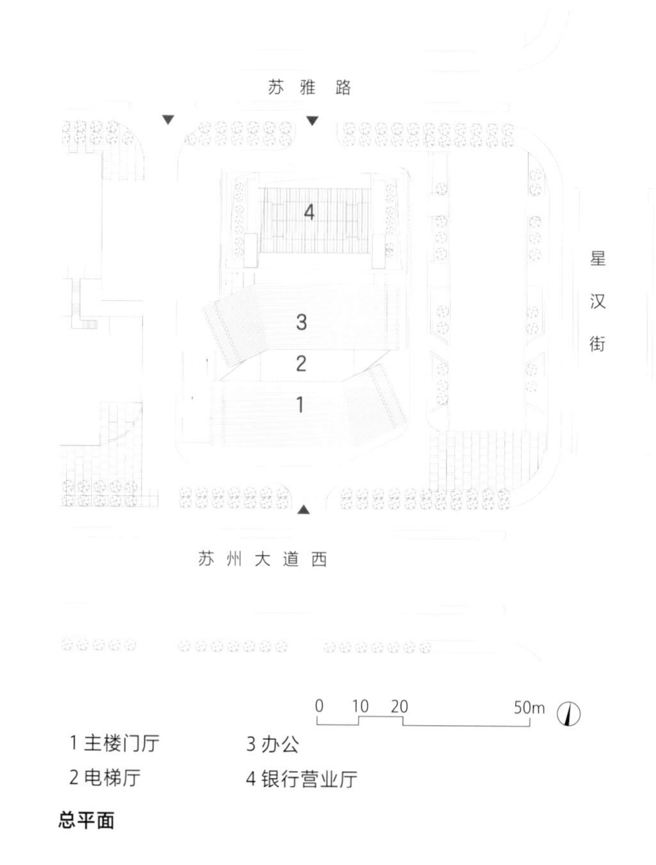

总平面

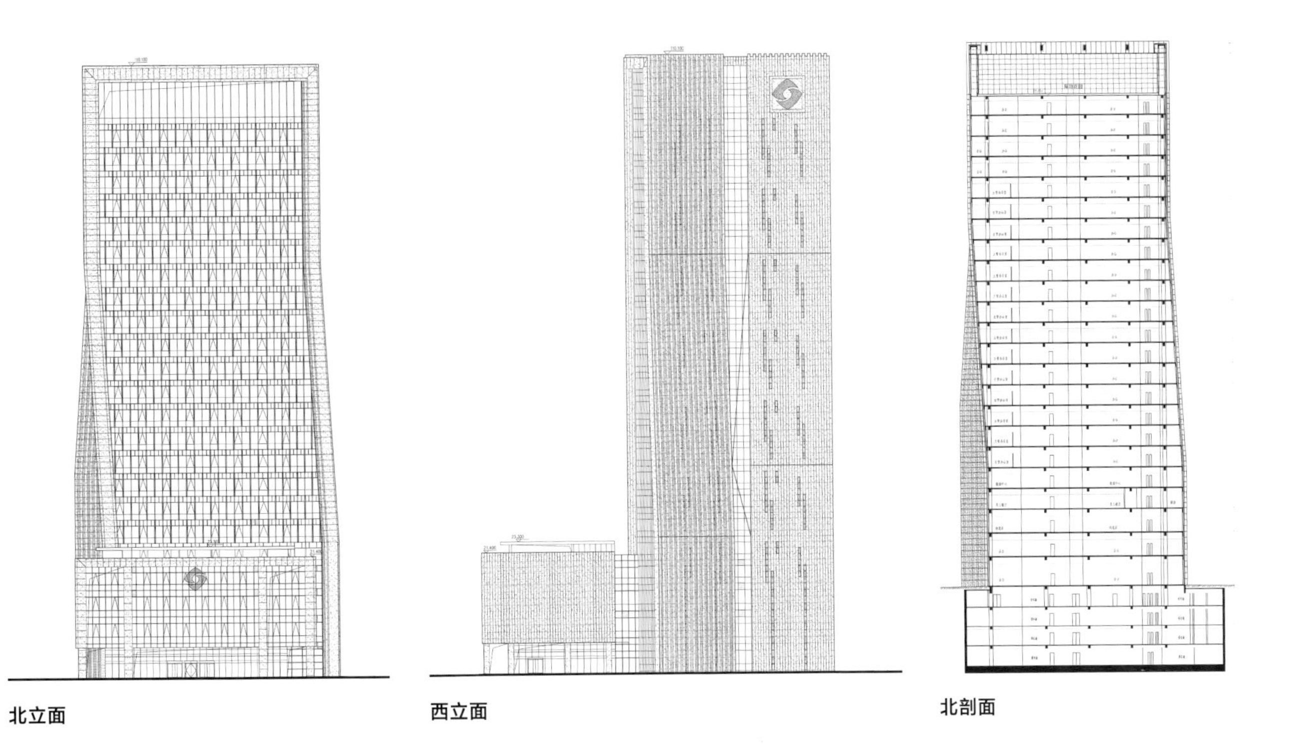

北立面　　西立面　　北剖面

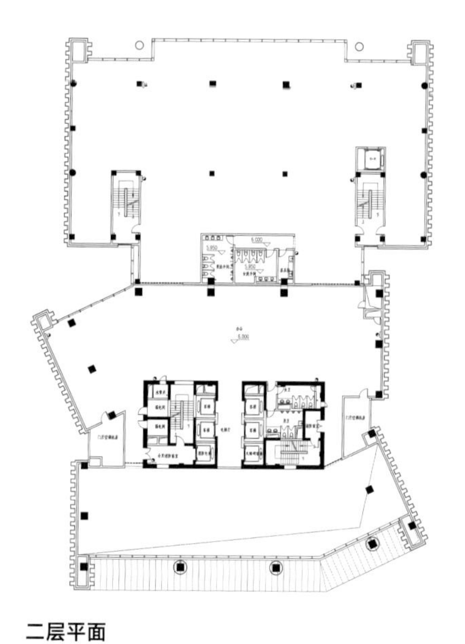

二层平面

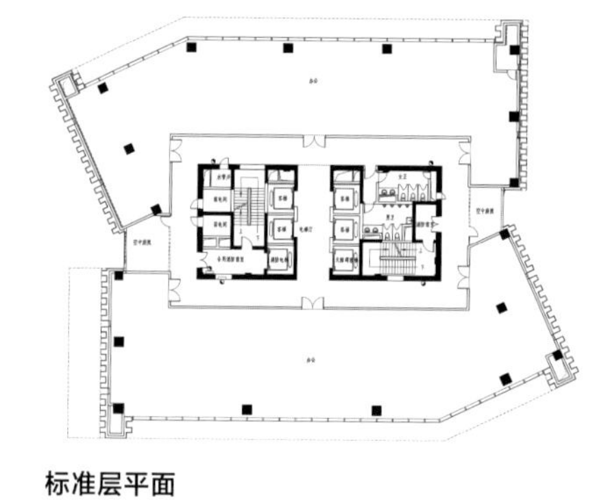

标准层平面

两个塔楼之间形成东西两个开放中庭，与南北立面的玻璃幕墙一起，体现建筑的通透性，通过对内部功能空间的强化，彰显“以客户为中心”的服务理念。生成后的体量很好地呼应周边环境，并通过虚实的对比，巧妙地体现银行建筑的“坚实性”和鼓励公众参与的“透明性”。

The two open atriums formed between the north and south volumes of the tower, along with the glass curtain walls of the north and south facades, reflect the building's permeability. By strengthening the internal functional space, the service concept of "customer-centric" is demonstrated. The generated volume responds aptly to the surrounding environment and skillfully reflects the "solidity" of the bank building and the "transparency" that encourages public participation through a comparison of virtual-real.

苏州国际博览中心

SUZHOU INTERNATIONAL EXPO CENTER

项目类型： 文体建筑　　**项目地点：** 江苏苏州
用地面积： 188600m^2　　**建筑面积：** 320000m^2
设计时间： 2003年　　**竣工时间：** 2014年
合作单位： SOM建筑设计事务所

第四届“科创杯”BIM大赛二等奖
2016年度江苏省第十七届优秀工程设计二等奖
2020年度江苏省城乡建设系统优秀勘察设计地下建筑及人防工程三等奖

苏州国际博览中心位于苏州工业园区金鸡湖畔，总体造型像一把打开的苏州折扇，既时尚又富有苏州婉约特色。扇片状的屋顶采用了不锈钢和玻璃材料，在阳光的照射下闪闪发光。扇形建筑体量与优美的前庭、流动的小河，以及景观花园、林荫大道融为一体。

Suzhou International Expo Center is located at the Jinji Lakeside in the Suzhou Industrial Park. Its overall shape is like an open Suzhou folding fan, which is fashionable and full of Suzhou graceful characteristics. The roof, which looks like folding fan blades, are made of stainless steel and glass, shining brightly in the sunlight. The fan-shaped building volumes are integrated with graceful front gardens and flowing creeks, as well as landscaped gardens and boulevards.

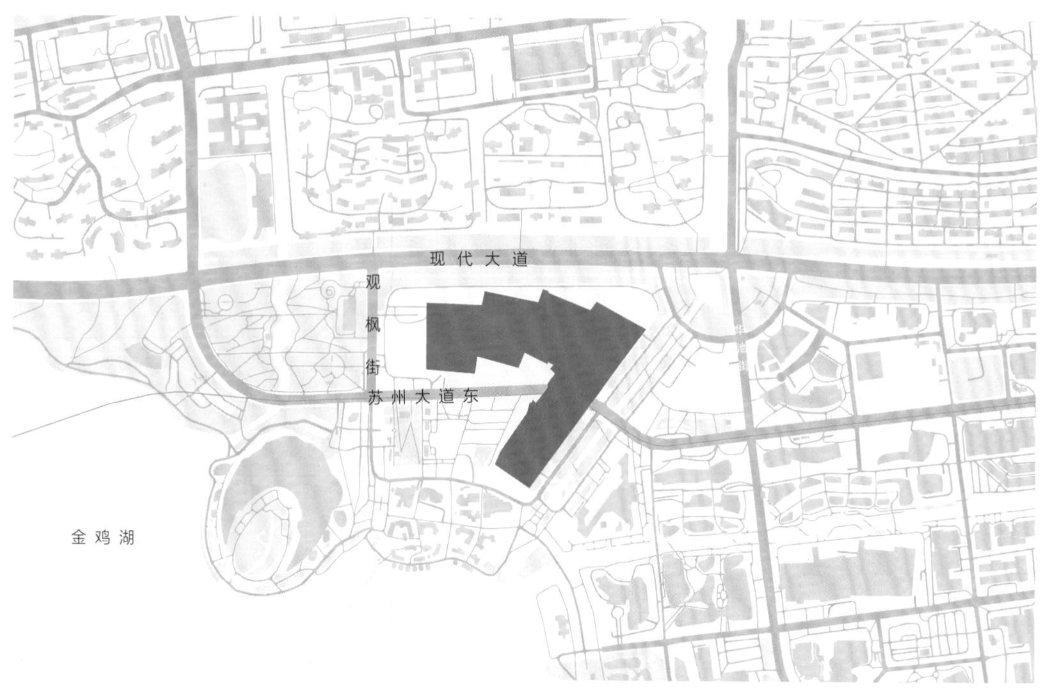

区位

设计将酒店和会议以水平分区和垂直分区相结合的方式进行功能叠加，满足会议人流和酒店住宿人流的最优化需求。在功能布局上，二号建筑底层为展厅，二层为8000m^{2}无柱宴会厅；一号建筑西侧部分为酒店区域，东侧为会议办公区，在垂直方向由开放过渡到私密，由大空间到小空间，尤其是客房区与办公区围合的内庭院，闹中取静。

The design combines the hotel and conference with horizontal and vertical zoning for a functional overlay to meet the optimized needs of the conference and hotel occupancy flows. In terms of functional layout, the ground floor of Building No. 2 is an exhibition hall, and the second floor is an 8000-square-meter pillar-less ballroom; the west side of Building No. 1 is partly a hotel area, and the east side is a conference and office area, which is vertically transitioned from openness to privacy, and from a large space to a small space, especially the inner courtyard surrounded by the guest room area and the office area, which is a quiet and peaceful place in the midst of a bustling city.

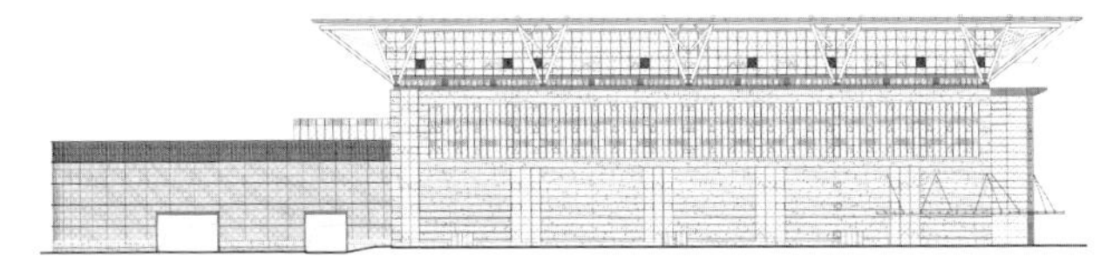

立面一

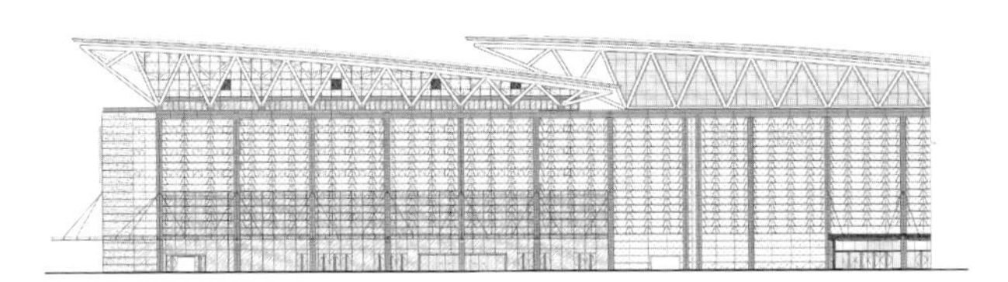

立面二

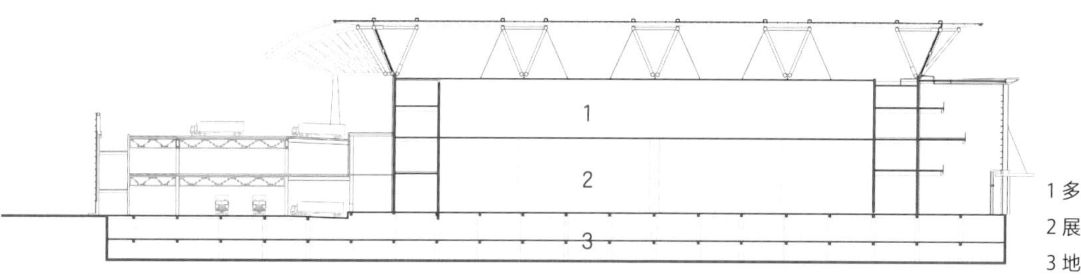

剖面

HYATT REGENCY

苏州科技馆与工业展览馆

SUZHOU SCIENCE HALL AND INDUSTRIAL EXHIBITION HALL

项目类型： 文体建筑　　**项目地点：** 江苏苏州
用地面积： 44036m²　　**建筑面积：** 61285m²
设计时间： 2020年　　**竣工时间：** 在建
合作单位： 帕金斯威尔建筑设计咨询（上海）有限公司

项目位于狮山广场北侧，与中部的剧场和南侧的博物馆一起，共同打造苏州人文环境新名片。设计尊重场地与狮山之间的关系，充分利用基地的自然环境优势。展馆建筑被设计构想成为狮山地形的视觉延伸部分，通过露台、室外屋顶花园和建筑总体形态等将场地与建筑巧妙融合。建筑从自然生态与传统文化中汲取灵感，展现现代化、前瞻性的建筑风貌，目的是创造具有历史长期性、功能全面性和视觉标志性的建筑，以此彰显创新、探索与发现的科技精神。

The project is located on the north side of Shishan Culture Square, together with the theater in the middle and the museum on the south side, to jointly create a new business card for Suzhou's cultural environment. The design respects the relationship between the site and Shishan, taking full advantage of the natural environment of the base. The exhibition hall is designed as a visual extension of the Shishan terrain, the site and the building are delicately integrated through the terrace, outdoor Roof garden, the overall form of the building, etc. Architecture draws inspiration from natural ecology and traditional culture, showcasing a modern and forward-looking architectural style. The purpose is to create buildings with long-term history, comprehensive functionality, and visual landmarks, demonstrating the spirit of innovation, exploration, and discovery.

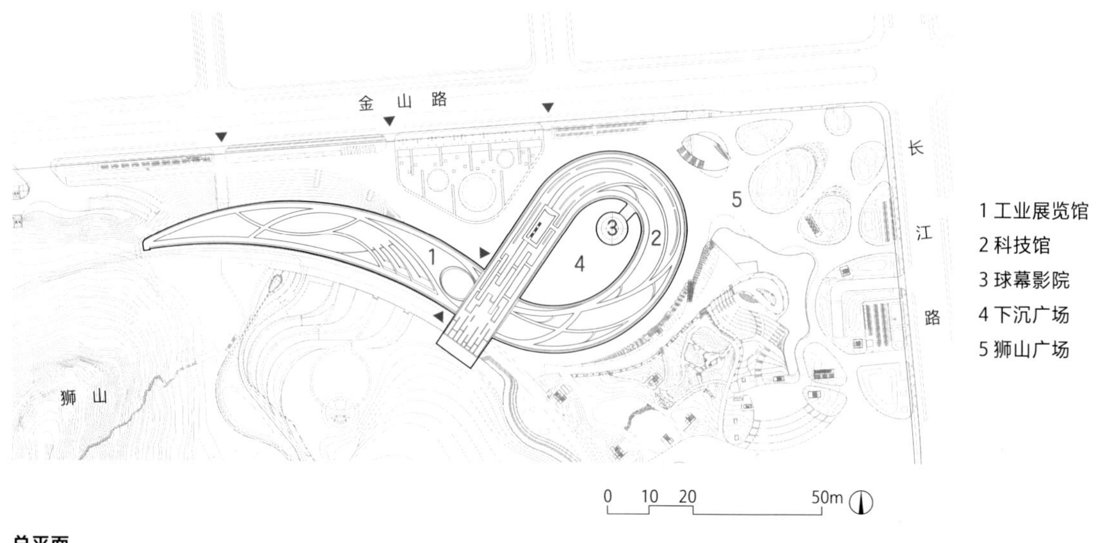

总平面

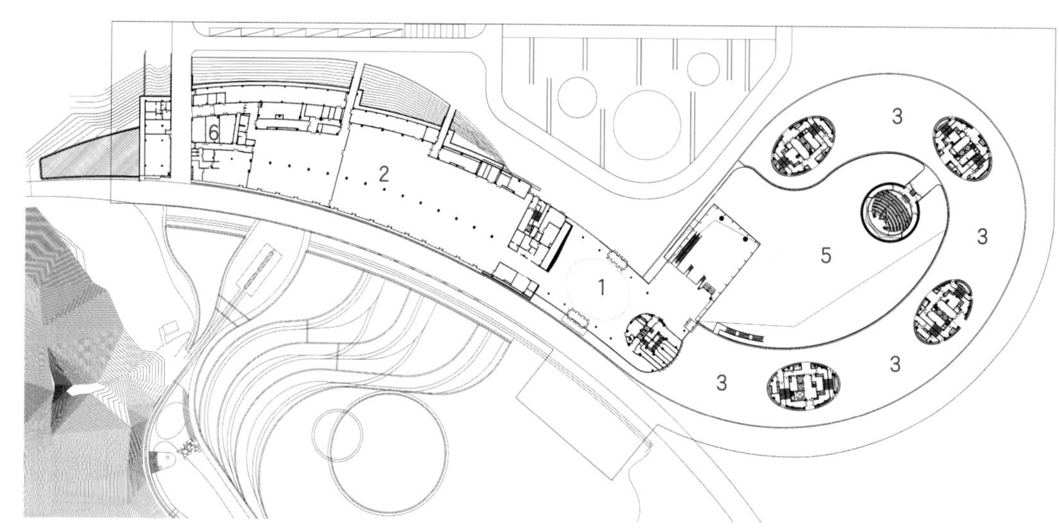

1 入口门厅
2 工业展厅
3 室外广场
4 球幕影院
5 室外下沉广场上空
6 设备用房区

一层平面

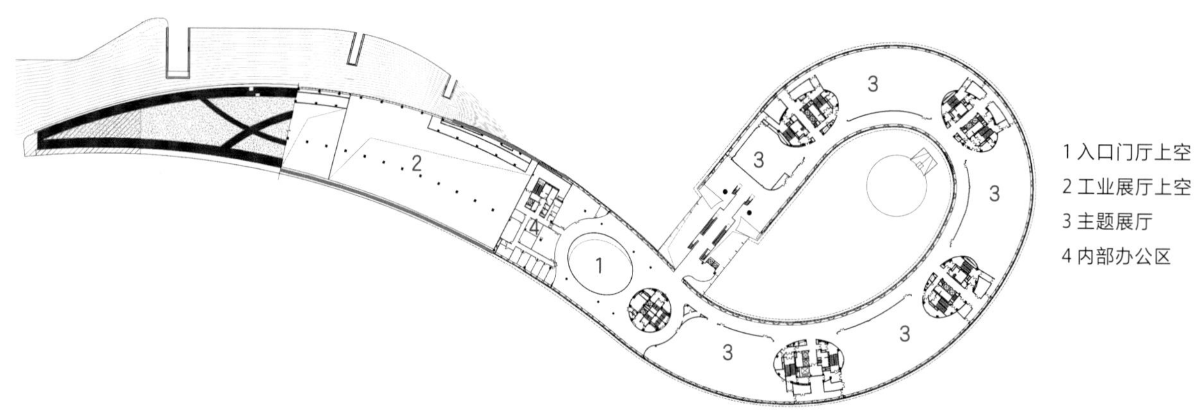

1 入口门厅上空
2 工业展厅上空
3 主题展厅
4 内部办公区

二层平面

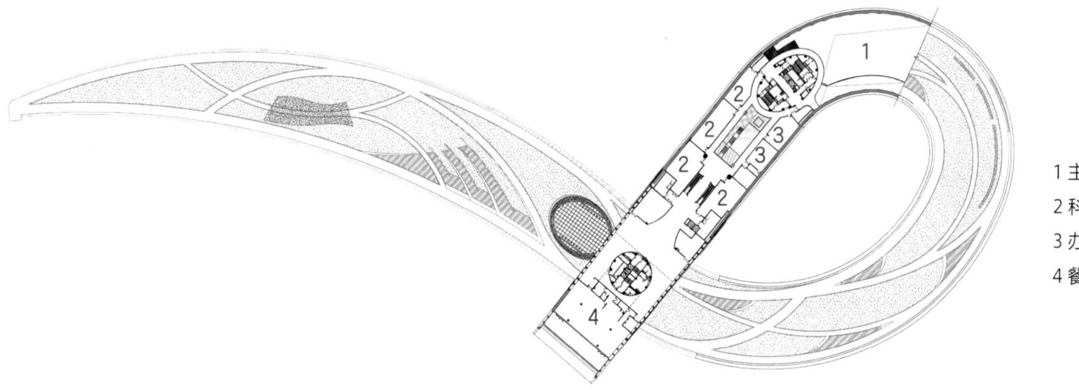

1 主题展厅上空
2 科创活动室
3 办公区
4 餐厅

三层平面

立面展开图

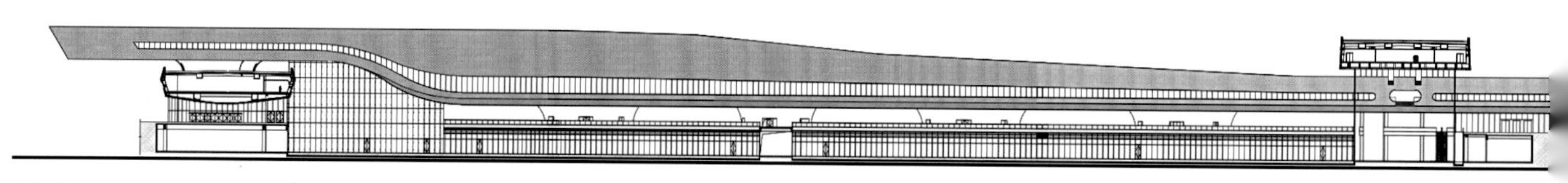

剖面展开图

项目遵循“山水融合”的设计理念，以中式的方法和手段来表达山与湖的融合，通过自然元素与科技创新相结合来发展设计概念。延绵的山体转变为一道弯曲折转的条形体，曲形的建筑形态悬浮在地面上方，摇曳攀升又回转而下，重新投入山的怀抱。由此生成的新展馆建筑形态与狮山及水岸区域构成了一个无限循环符号，隐喻了科技进步与自然奇观之间不可阻断的内在联系。

The project follows the design concept of "landscape city", expressing the fusion of mountains and lakes through Chinese methods, and developing the design concept through the combination of natural elements and technological innovation. The rolling mountains quickly transforms into a curved and twisted strip shape. The curved architectural form hang in the air, sways and climbs, then turns back down and rejoins the embrace of the mountain. The resulting architectural form of the new exhibition hall forms an infinite circular symbol with the Shishan and lakeside area, symbolizing the unbreakable intrinsic link between technological progress and natural wonders.

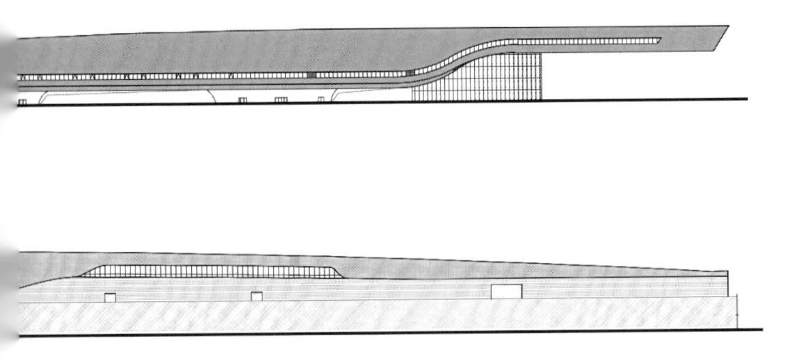

苏州橙天360剧场

SUZHOU OSGH 360 THEATER

项目类型：艺术中心　　**项目地点**：江苏苏州

用地面积：33870m²　　**建筑面积**：34939m²

设计时间：2019年　　**竣工时间**：2022年

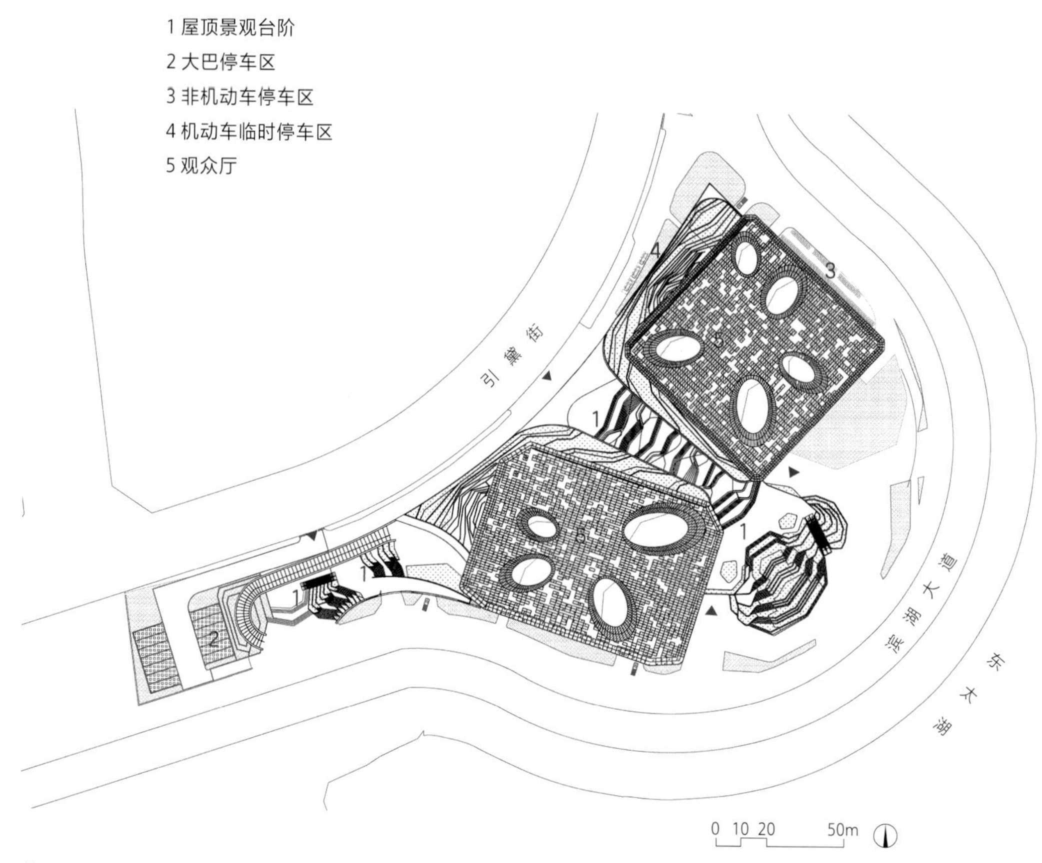

总平面

项目位于吴中太湖苏州湾北岸，地处吴中太湖新城核心区，是太湖周边重要的文化标志性建筑，着力打造集观演与文化休闲于一体的文化娱乐中心。设计采用“湖畔砾石”的理念，坐落在湖边的建筑宛如自然环境中生长出来的石头，坡地景观的柔美自由和两块砾石的硬朗干脆形成鲜明对比，突显了本项目与太湖文化及生态环境协调一致的理念。

The project is located on the north shore of Suzhou Bay, Taihu New Town, the core area of Wuzhong. It is an important cultural landmark around the Taihu Lake, and focuses on creating a cultural and entertainment center that integrates performance watching and cultural recreation. The design adopts the concept of "lakeside gravel", where the buildings are located on the lake like stones growing out of the natural environment. The natural softness of the sloping landscape contrasts with the hardness and crispness of the two "gravels", emphasizing the concept of harmonizing the project with the culture and ecological environment of Taihu.

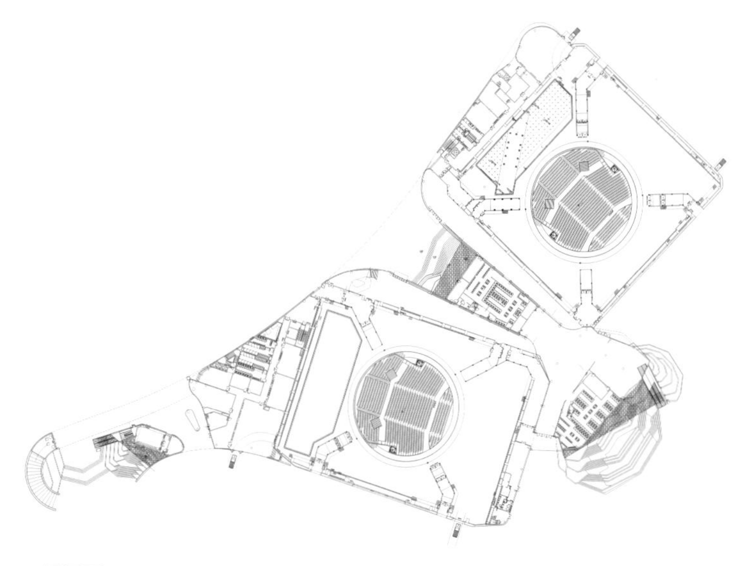

一层平面

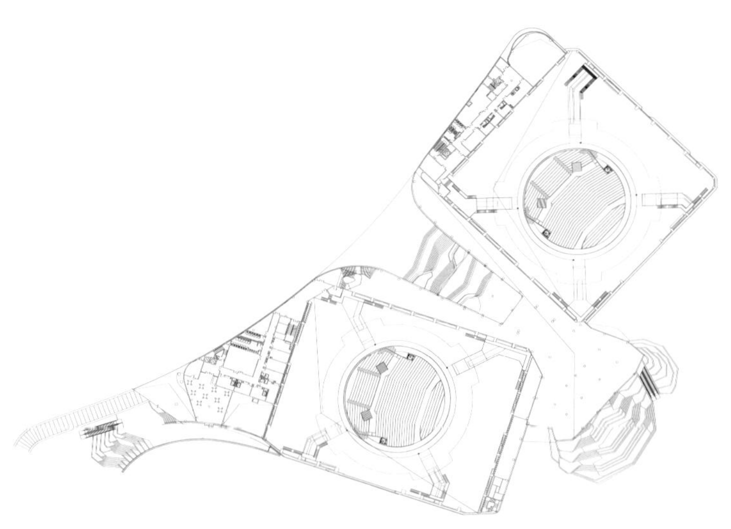

二层平面

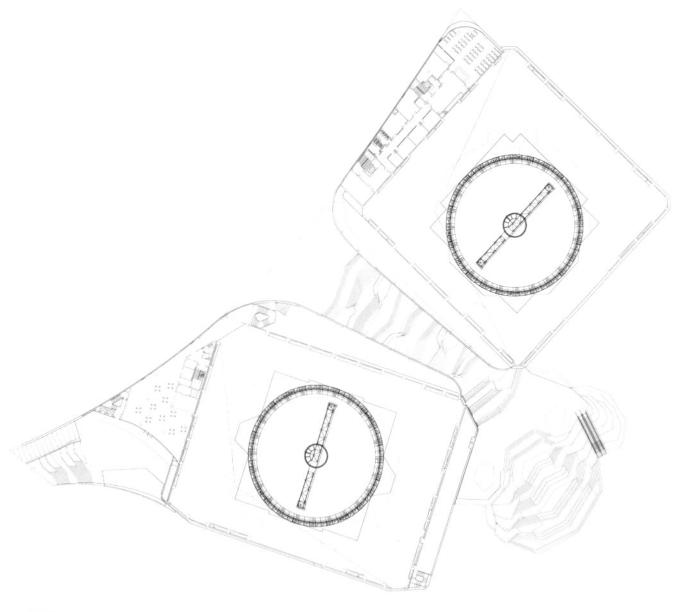

三层平面

项目周边均为滨水开放空间，因此设计将整个区域打造成立体的生态公园，配合观演和游玩两条功能流线，室内室外互相连通，形成完整的湖边文化旅游业态链，不仅将整体的生态绿线进行延续，还营造出生动有趣的水岸游乐体验，让自然元素和城市空间在此完美结合。

The project is surrounded by waterfront open space, so the design will make the whole area into a three-dimensional ecological park, with two functional streams of performance and play, indoor and outdoor interconnected, forming a complete lakeside cultural tourism chain, not only to continue the overall ecological green line, but also to create a vivid and interesting lakeside play experience, so that the natural elements and urban space in this perfect combination.

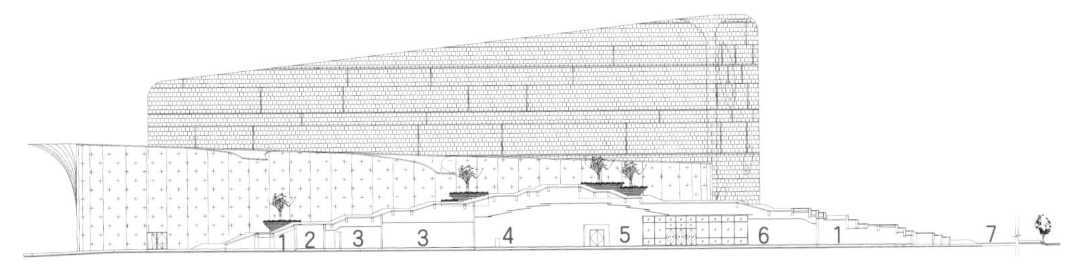

1 结构空腔（室外）
2 储物间
3 卫生间
4 服务台
5 公共空间
6 咖啡厅
7 室外（临湖广场）
8 剧场
9 公共空间
10 机房
11 避难走道

典型剖面一

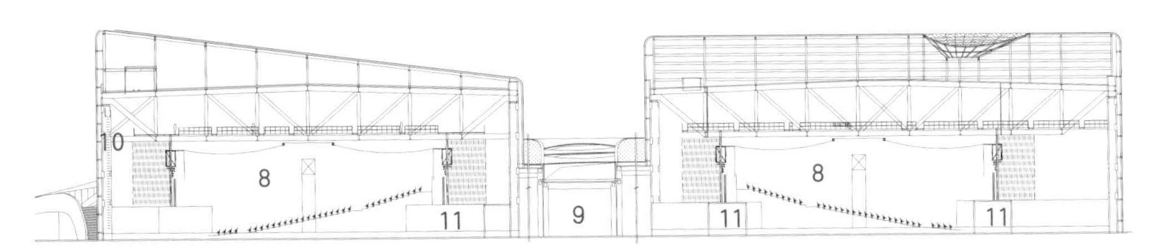

典型剖面二

江苏硅谷展示馆

JIANGSU SILICON VALLEY EXHIBITION HALL

项目类型：展示中心　　**项目地点**：江苏句容

用地面积：1949m^2　　**建筑面积**：4461m^2

设计时间：2017年　　**竣工时间**：2019年

2020年度江苏省第十九届优秀工程设计一等奖
2020年度江苏省城乡建设系统优秀勘察设计一等奖
2018年度江苏省土木建筑学会第十二届“建筑创作奖”一等奖

项目位于句容市江苏硅谷科技园，东侧引入园区中轴线，西侧紧邻自然湿地，是科技园区的视觉及功能核心，也是句容境内宁容快速路旁的标志性建筑物。建筑形体由三片立体旋转上升的叶子及中间的裙体组合而成。内部三层展示及办公空间围绕核心中庭布置，中庭内的无梁柱钢楼梯，是内部交通与视线、技术与美学结合的仪式焦点。

The project is located in Jiangsu Silicon Valley Science and Technology Park, Jurong, introducing the central axis of the park on the east side and adjacent to the natural wetland on the west side, which is the visual and functional core of the Science and Technology Park, and also a landmark building beside Ningyong Expressway within Jurong. The overall shape of the building is composed of three three-dimensional rotating and rising leaves and a central skirt. Three levels of internal exhibition and office space are arranged around a core atrium with a beamless and column-free steel staircase, a ceremonial focal point for the combination of internal traffic and sightlines, technology and aesthetics.

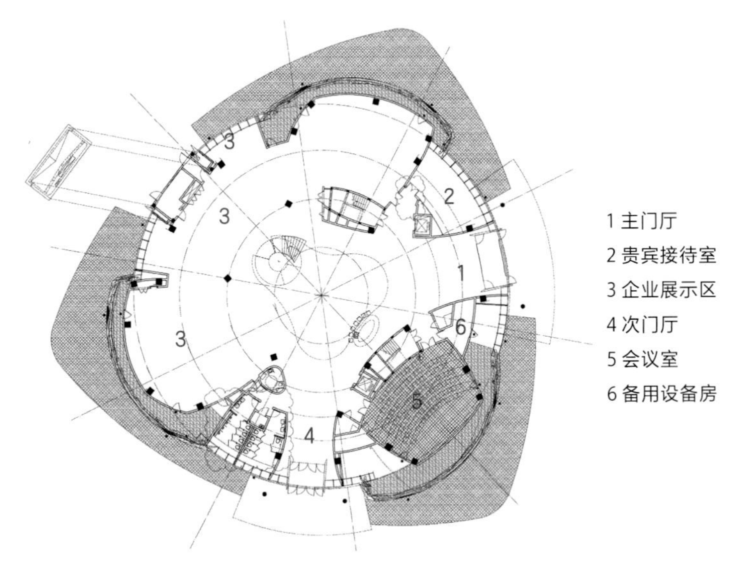

1 主门厅
2 贵宾接待室
3 企业展示区
4 次门厅
5 会议室
6 备用设备房

一层平面

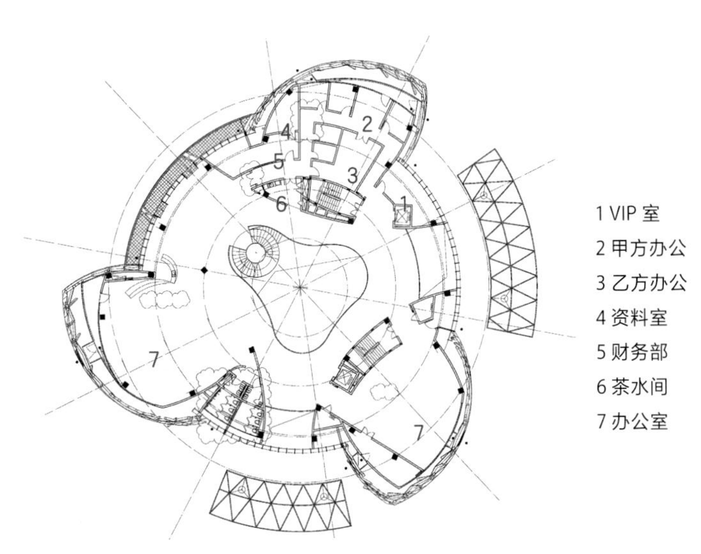

1 VIP 室
2 甲方办公
3 乙方办公
4 资料室
5 财务部
6 茶水间
7 办公室

二层平面

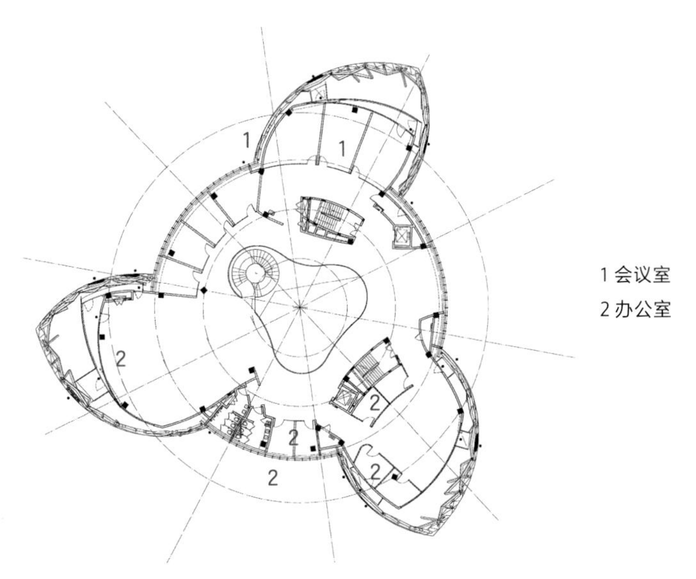

1 会议室
2 办公室

三层平面

项目以EPC模式进行设计建造组织，建筑造型复杂，土建、机电、钢结构、幕墙需要高度协同，全程采用数字化技术手段支撑各项工作，如方案阶段参数造型、方案深化及施工图阶段的BIM正向设计，施工阶段的施工模拟、点云扫描复核等。通过数字技术手段的应用，在设计理念落地及成本控制上均取得良好的效果。

The project is organized in EPC mode for design and construction, with complex shapes and a high degree of synergy among civil engineering, mechanical and electrical, steel structure and curtain wall. The whole process adopts digital technology to support the work, such as parametric modeling and scheme deepening in the scheme stage, BIM forward design in the construction drawing stage, construction simulation and point cloud scanning review in the construction stage. Through the application of digital technology methods, good results have been achieved in the design concept and cost control.

BIM 模型图

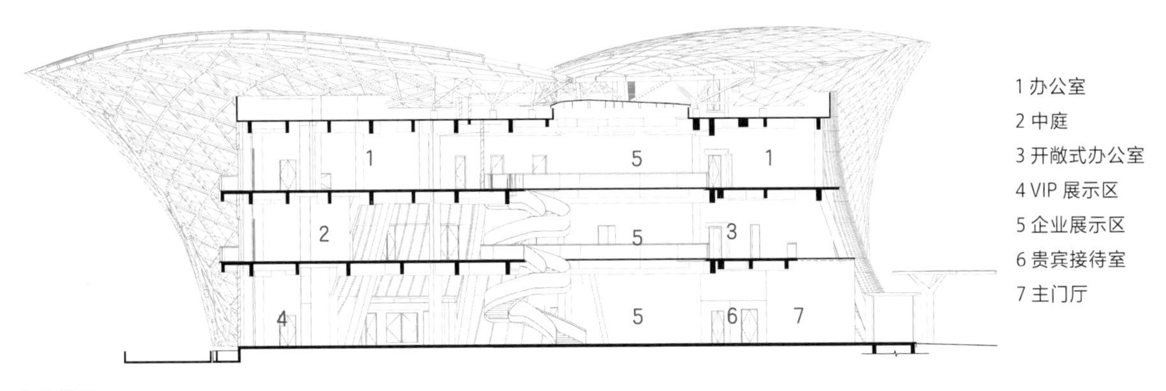
1 办公室
2 中庭
3 开敞式办公室
4 VIP 展示区
5 企业展示区
6 贵宾接待室
7 主门厅

1-1 剖面

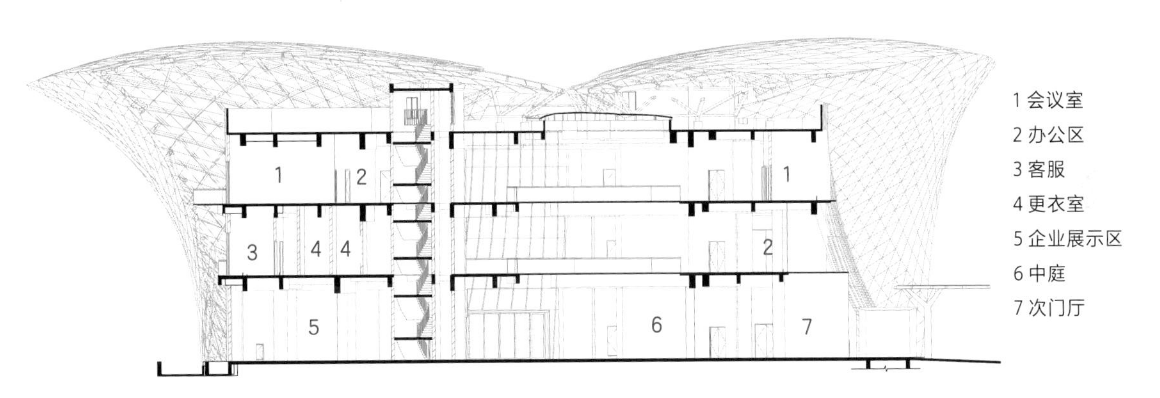
1 会议室
2 办公区
3 客服
4 更衣室
5 企业展示区
6 中庭
7 次门厅

2-2 剖面

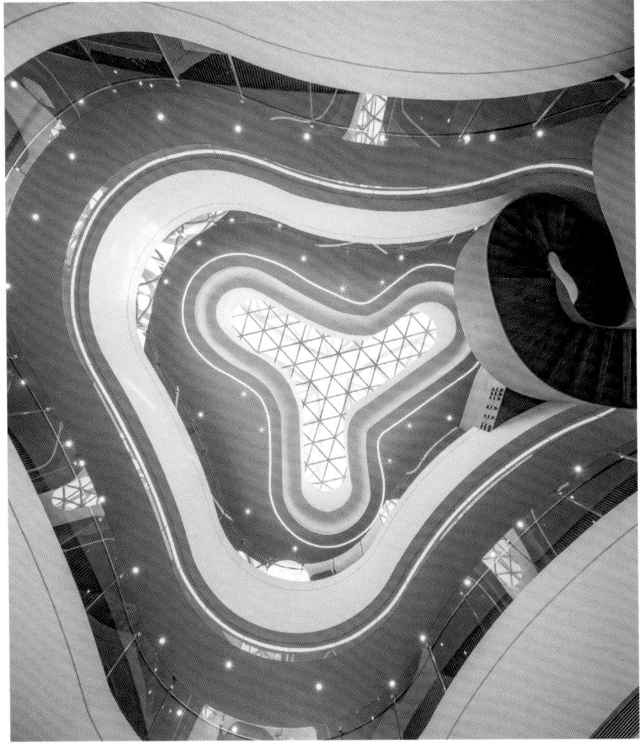

苏州工业园区北部文体中心

SIP NORTHERN CULTURE AND SPORTS CENTER

项目类型： 文体建筑　　**项目地点：** 江苏苏州
用地面积： 23830m²　　**建筑面积：** 48848m²
设计时间： 2017年　　**竣工时间：** 2021年

2022年度江苏省第二十届优秀工程设计一等奖
2022年度江苏省城乡建设系统优秀勘察设计一等奖
第五届AIIDA美国国际创新设计大奖文化建筑金奖

项目位于苏州工业园区，主要由剧院、综合体育馆、游泳馆、综合图书馆四大功能板块组成。四个场馆根据功能及定位分开设计，并且通过半开放式的平台连接，保证了每个场馆均有独立的出入口可以独立运营，减少不同场馆因使用时间不同而造成的能源损耗。

项目最大限度的对城市开放，各功能体量之间尽可能多的与自然结合，通过屋顶运动场、水院、绿化广场、共享平台、漫游步道等开放空间，融入城市肌理。设计用宜人的公共空间尺度，打造一座亲民的公共服务中心，用多元化的立面及功能，不断激发城市的创新活力，满足市民在文化、商业、娱乐、休闲等多方面的诉求，服务社会、成为城市中充满活力的公共场所。

区位

Located in Suzhou Industrial Park, the project is mainly composed of four functional sections: theater, comprehensive gymnasium, swimming pool and comprehensive library. The four venues are designed separately according to their function and orientation, and are connected by semi-open platforms. It ensures that each venue has independent entrances and exits and can operate independently, reducing the energy loss caused by different venues due to different usage times.

The project maximizes openness to the city, with as much integration with nature as possible between the various functional volumes. The building is integrated into the urban fabric through open spaces such as rooftop playgrounds, water courtyards, green plazas, shared platforms, and rambling walkways. The building design uses a pleasant scale of public space to create a citizen-friendly public integrated service center; with diversified facades and functions, it will continue to stimulate the city's innovation and vitality, satisfy the public's demands in culture, commerce, entertainment, leisure and other aspects, serve the society and become a vibrant public place in the city.

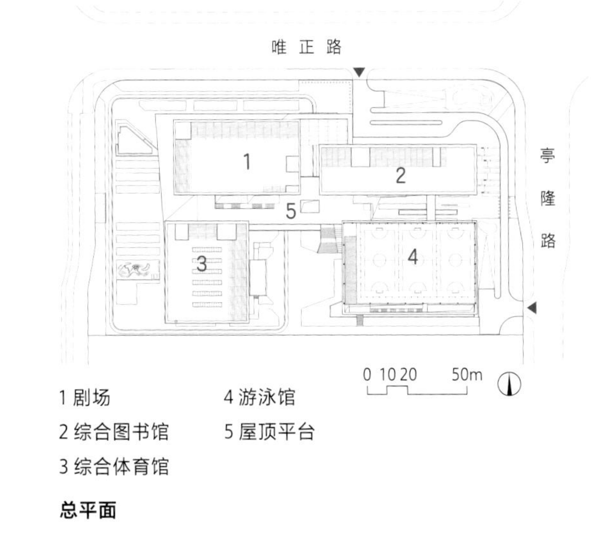

总平面

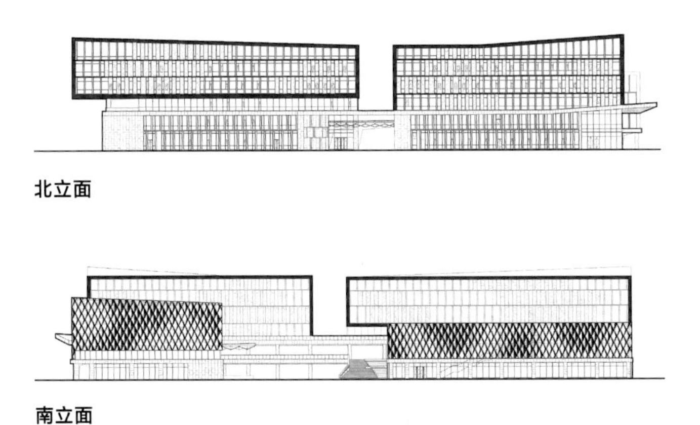
北立面

南立面

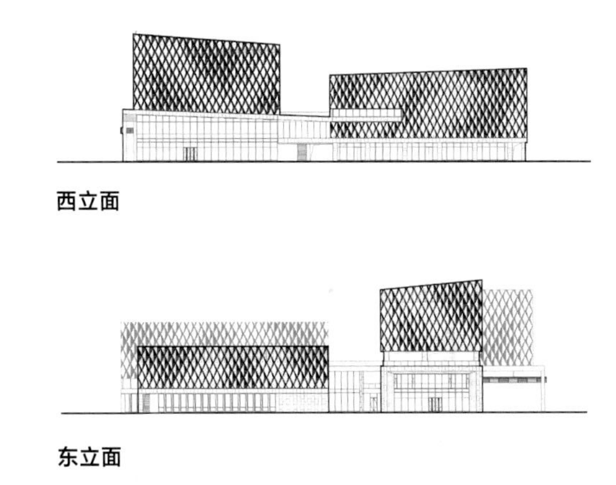
西立面

东立面

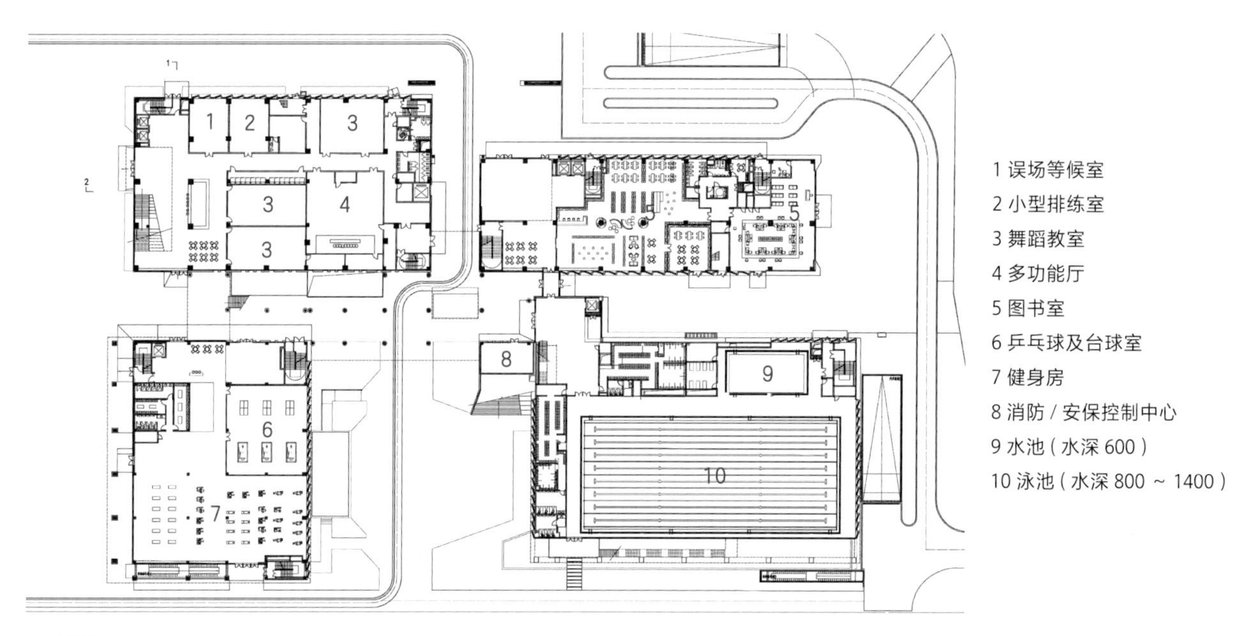
1 误场等候室
2 小型排练室
3 舞蹈教室
4 多功能厅
5 图书室
6 乒乓球及台球室
7 健身房
8 消防 / 安保控制中心
9 水池（水深 600）
10 泳池（水深 800 ~ 1400）

一层平面

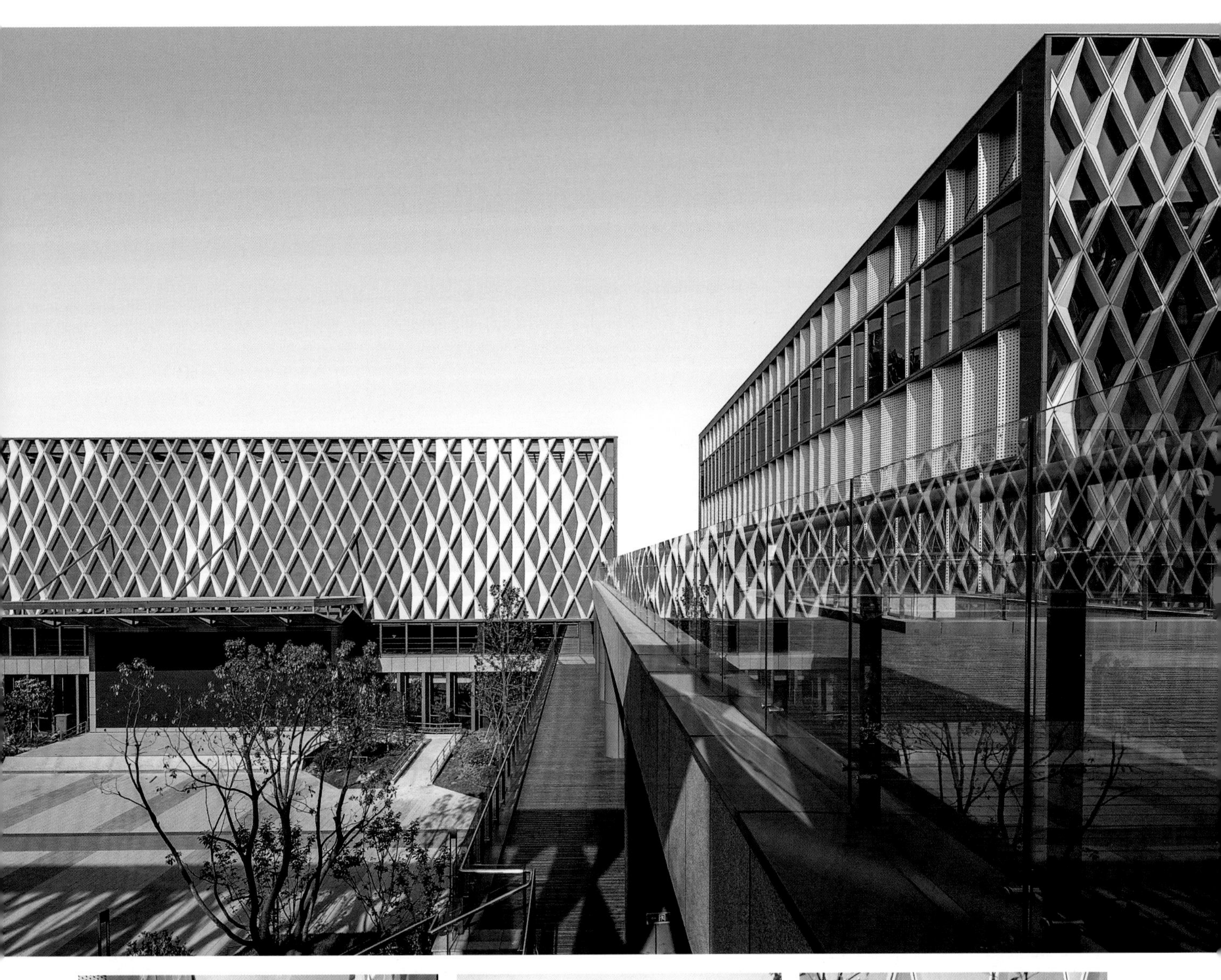

潘祖荫故居改造

RECONSTRUCTION OF PAN ZUYIN'S FORMER RESIDENCE

项目类型：既有建筑改造　　**项目地点**：江苏苏州
用地面积：2935m²　　**建筑面积**：3158m²
设计时间：2011年　　**竣工时间**：2019年

2015年度第五届“华彩奖”金奖
2015年度全国优秀工程勘察设计行业奖二等奖
2014年度江苏省第十六届优秀工程设计一等奖
2014年度江苏省城乡建设系统优秀勘察设计一等奖
2021年度江苏省城乡建设系统优秀勘察设计二等奖
2021年度江苏省优秀工程勘察设计行业奖一等奖

平江历史街区中有大量大户宅院，历史文化遗存众多，潘祖荫故居就是其中一处。故居又名竹山堂，是一座三路五进、坐北朝南的大宅子，由清末探花潘祖荫晚年退居苏州时改建住所而来。这里曾被用作工厂、招待所，历经岁月的沧桑，昔日盛景不再，宅第开始变得破败。为了延续建筑文脉，保护名士故居，潘宅作为苏州市控制保护建筑进行维修整治。

设计坚持“修旧如旧”的原则，参考史料记载，拆除庭院搭建，将东路园林还原，恢复曾经大户宅园的城市肌理，最大限度地复原建筑原有的风貌与格局，力争与整个历史街区的气息氛围融为一体。

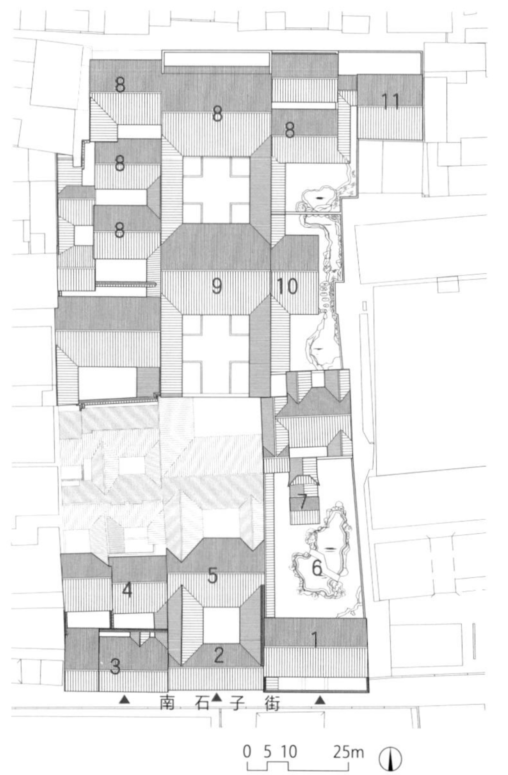

1 花厅　5 会议厅　9 内厅
2 探花书房　6 东花园　10 餐饮区
3 沙龙　7 船舫　11 厨房
4 多功能厅　8 客房区

总平面

The historic district of Pingjiang has a large number of large houses and many historical and cultural relics, one of which is the Pan Zuyin's Former Residence. The Pan Zuyin's Former Residence, also known as Zhusantang, is a three-way, five-entry, south-facing mansion that was rebuilt from the residence of Pan Zuyin, the Tanhua* in late Qing Dynasty, when he retired to Suzhou in his twilight years. It was once used as a factory and a guesthouse, and after the vicissitudes of time, the former splendor was no longer there, and the mansion began to fall into disrepair. In order to continue the architectural context and protect the former residence of the famous scholar, the Pan's Residence was repaired and improved as a protected building in Suzhou.

The design adheres to the principle of "repair the old as before" and refer to historical records. The courtyards were dismantled and erected, the East Road gardens were restored, and the urban texture of what was once a large house was restored. The residence have been restored to their original appearance and layout to the greatest extent possible, in an effort to blend in with the atmosphere of the entire historic district.

* Tanhua: number three in national civil examinations in feudal China.

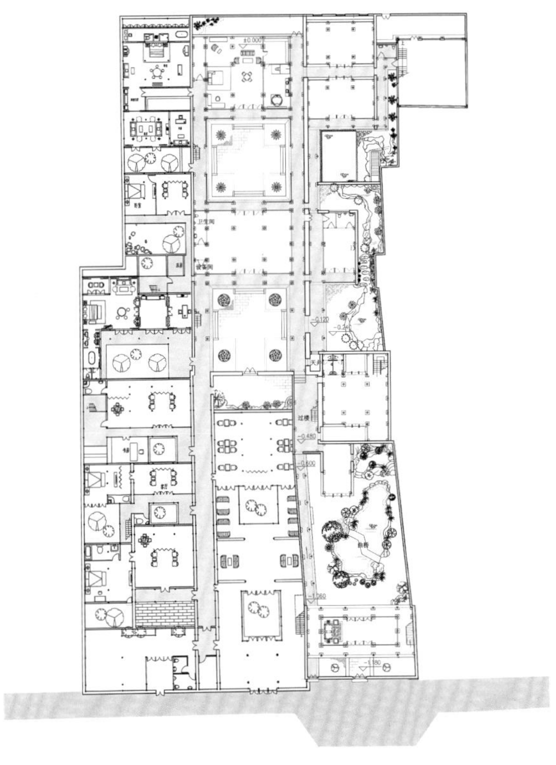

一层平面

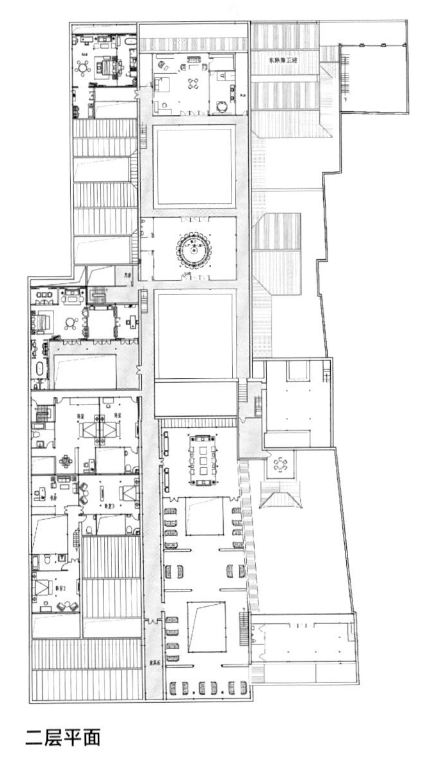

二层平面

改造前（下）后（上）对比

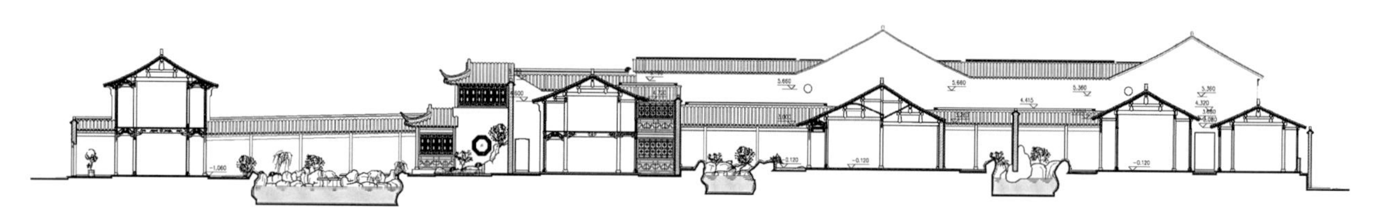

东路中贴剖面

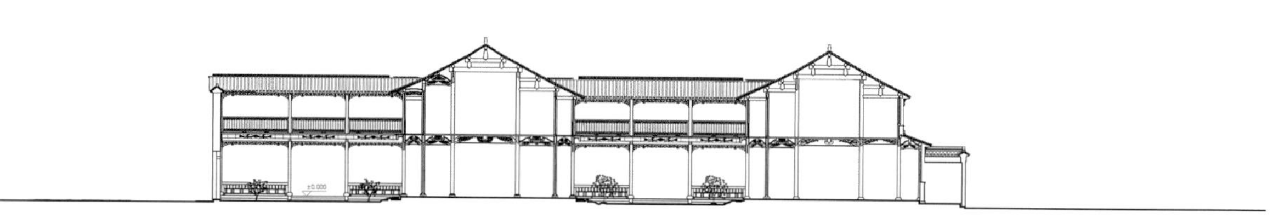

中路中贴剖面

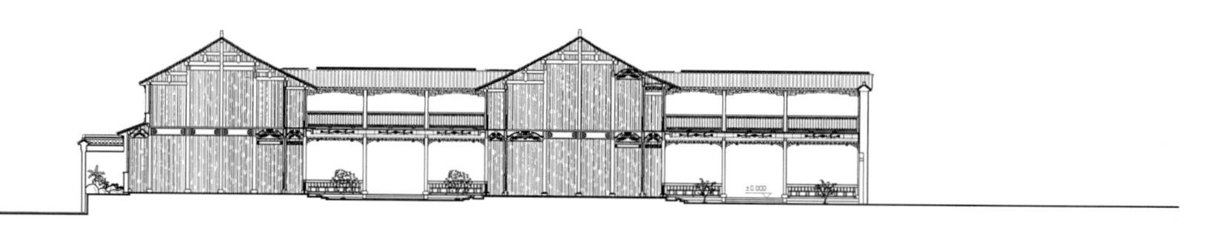

边路中贴剖面

潘祖荫故居定位为精品酒店，不仅符合古建筑从“宅院”到“客房”的同质性功能置换，对原建筑的影响降到最低。同时对于拉动平江路由“线”向“面”的发展，带动整个片区的经济价值也十分有益。潘祖荫本人的历史生平、老宅的建筑特色与宅院文化、老宅的收藏文化等，都是将潘祖荫故居改造成精品酒店的特色文化主题，也是人们在其中体验的重点。

The positioning of Pan Zuyin's Former Residence as a boutique hotel is not only in line with the homogeneous functional replacement of ancient buildings from "mansions" to "guest rooms", but also minimizes the impact on the original building. At the same time, it is also very useful in promoting the development of Pingjiang Road from "line" to "surface" and enhancing the economic value of the whole area. The historical life of Pan Zuyin, the architectural characteristics, the mansion culture and collection culture of the old house, are all the special cultural themes of the transformation of Pan Zuyin's Former Residence into a boutique hotel, and the key that people want to experience in it.

书香世家·平江府酒店

SCHOLARS HOTEL SUZHOU PINGJIANGFU

项目类型：既有建筑改造/酒店	**项目地点**：江苏苏州
用地面积：15798m^2	**建筑面积**：19233m^2
设计时间：2009年	**竣工时间**：2010年

2011年度全国优秀工程勘察设计传统建筑二等奖

2015年度华彩奖金奖

2014年度江苏省第十六届优秀工程设计一等奖

书香世家·平江府酒店坐落在苏州古城的保护中心。该项目以“北半园”为依托，将一群废弃的厂房和一个破旧的园林改造为融地域元素与现代生活于一体的五星级创意文化酒店。

Scholars Hotel Suzhou Pingjiangfu is located in the heart of Suzhou's ancient conservation center. The project is based on the "North Half Garden", transformed the abandoned factory and the dilapidated garden into a five-star creative and cultural hotel that combines the regional elements with modern luxury.

改造前总平面

改造后总平面

改造前（下）后（上）对比

通过对旧建筑、园林改造与整合前后的对比分析，探索在追寻旧建筑、旧园林历史记忆的同时如何满足精品酒店的功能，并将这段历史文化延续在酒店空间的各个角落。书香世家·平江府酒店以富有想象力的当代设计，融合当地风格及本土材质，将姑苏风韵尽情展现在墙壁的纹理中，当走在老旧地板上，享受着现代服务，空间氛围却是静谧而充满活力的。

Through the comparative analysis of the old buildings and gardens before and after the renovation and integration, it explores how to satisfy the functions of the boutique hotel while tracing the historical memories of the old buildings and gardens, and how to perpetuate this history and culture in all corners of the hotel space.Scholars Hotel Suzhou Pingjiangfu's imaginative contemporary design blends local styles and materials to bring out the charm of Suzhou in the textures of the walls, while walking on old floors and enjoying modern services, the atmosphere of the space is serene and vibrant.

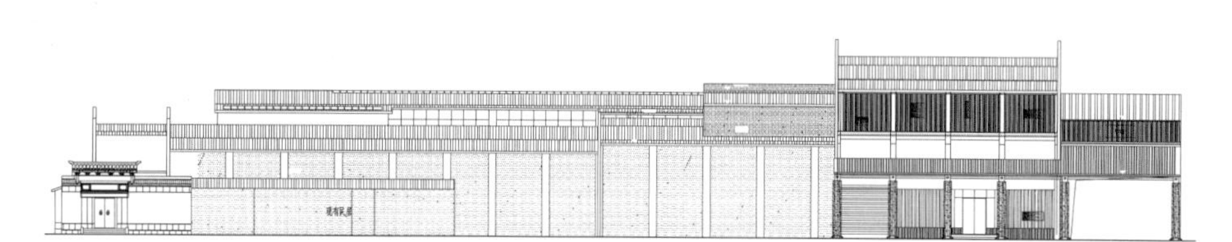

八、九号楼南立面

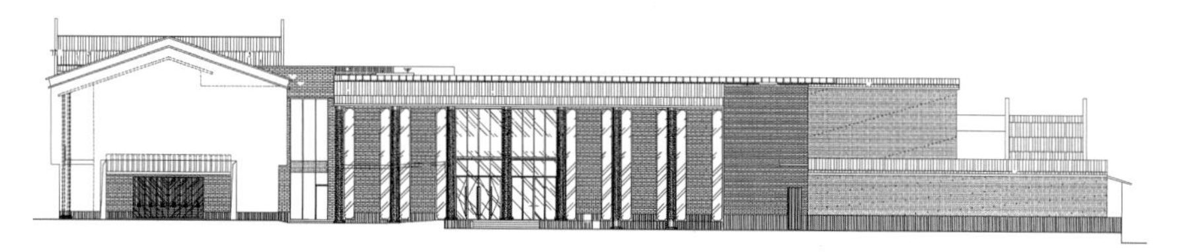

八、九号楼北立面

知足軒

苏州太美·逸郡酒店

G-LUXE BY GLORIA SUZHOU TAIMEI

项目类型：酒店建筑　　**项目地点**：江苏苏州

用地面积：47877m^2　　**建筑面积**：20935m^2

设计时间：2016 年　　**竣工时间**：2019年

2022年度江苏省第二十届优秀工程设计一等奖
2021年江苏省城乡建设系统优秀勘察设计一等奖
2022年度江苏省优秀工程勘察设计行业奖装饰一等奖
2020年度江苏省土木建筑学会第十四届“建筑创作奖”一等奖
第八届“创新杯”建筑信息模型（BIM）应用大赛BIM普及应用奖
第四届日本IDPA国际先锋设计大奖建筑银奖
第四届日本IDPA国际先锋设计大奖室内银奖
第五届AIIDA美国国际创新设计大奖国际创新奖
2021 Active House Award三等奖

1 公共区
2 高区客房
3 北区客房
4 无边泳池
5 景观绿地

总平面

项目地处太湖度假区渔洋山风景区，位于舟山路与环太湖大道三岔路口西北角，南侧紧邻环太湖大道与苏州海洋馆。项目背倚渔洋山，面向太湖，是风水绝佳之处。酒店依山而建，山麓层层叠起，将太湖风景呈现于视野之中，苏州太湖山水之秀美，此处一览无余。建筑之美，在于自然景观与艺术表现的结合。

酒店以优美的太湖生态为背景，提炼吴文化、太湖文化、渔洋山文化的精华，汲取苏州本地传统艺术为设计灵感，向世人呈现一幅风景秀美、物产丰富、诗情画意、充满人文气息的太湖图卷。

The project is located at the northwest corner of the three-way intersection of Zhoushan Road and Huan Taihu Avenue, Yuyangshan Scenic Area of Taihu Resort District, and adjacent to Huan Taihu Avenue and Suzhou Oceanarium on the south side. The project is backed by Yuyangshan and facing Taihu, which is an excellent place for feng shui. The hotel is built on the mountain, the foothills of which rise one after another, presenting the Taihu scenery in its view. The beauty of the Taihu landscape is visible here. The beauty of architecture lies in the combination of natural landscape and artistic expression.

With the beautiful ecology of Taihu as the background, the hotel refines the essence of Wu culture, Taihu culture and Yuyangshan culture, and draws on the local traditional art of Suzhou as the design inspiration, presenting to the world a picture scroll of Taihu with beautiful scenery, rich products, poetic feelings and full of humanistic atmosphere.

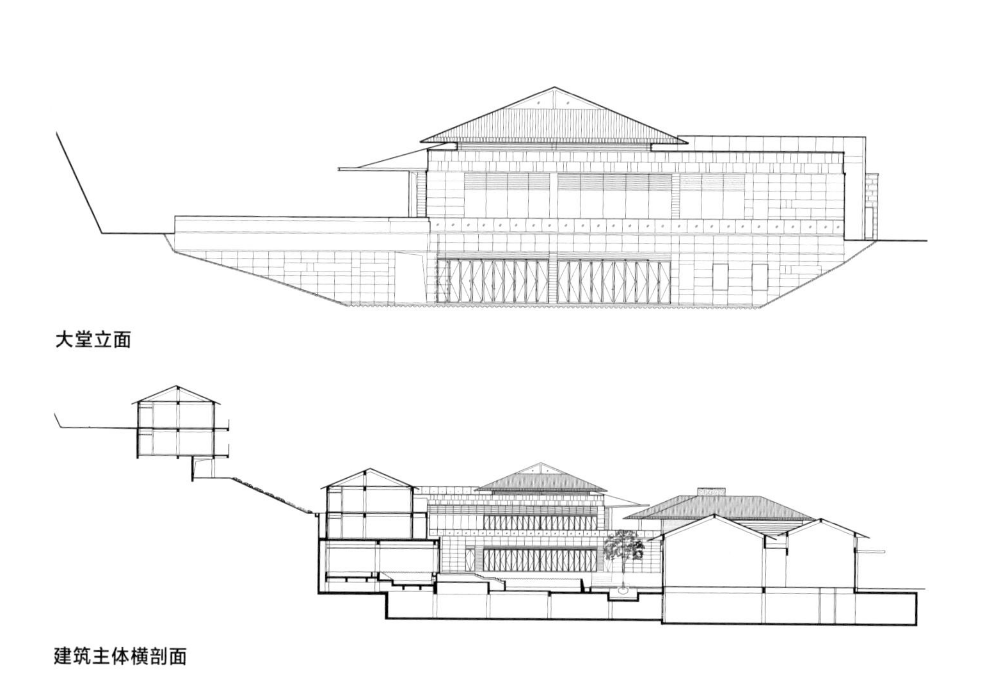

大堂立面

建筑主体横剖面

方案设计考虑含蓄而隐逸的东方文人精神与温良娴雅的生活姿态，力图打造恬淡精致的度假体验。同时，基地毗邻太湖，太湖呈现开阔的水平向景观也是重要考量因素。场地设计上试图做到清静，回归自然，利用山谷本身丰富的空间体验来塑造景观，创造建筑、人、自然和谐共生的亲密关系。

The design takes into account the implicit and hidden spirit of the Oriental literati and the gentle and elegant lifestyle, and seeks to create a tranquil and refined vacation experience. Meanwhile, the fact that the base is adjacent to Taihu, which presents an open horizontal view, is an important consideration. The site design attempts to achieve tranquility and return to nature, utilizing the rich spatial experience of the valley itself to shape the landscape and create an intimate relationship of harmonious symbiosis between architecture, people and nature.

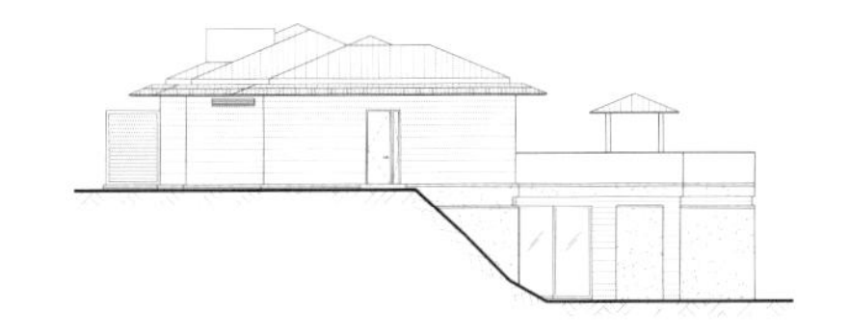

C 套型客房西立面

C 套型客房东立面

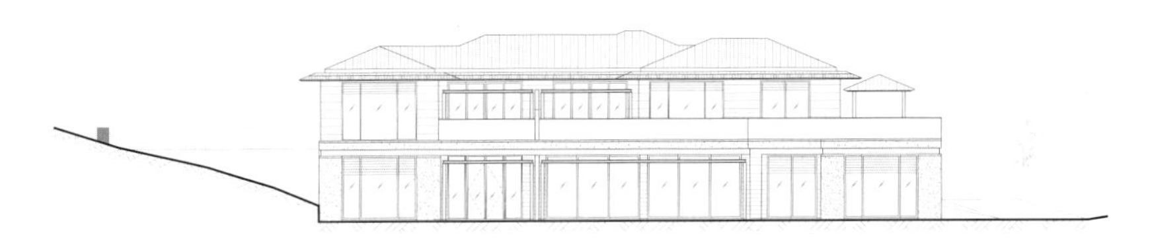

C 套型客房南立面

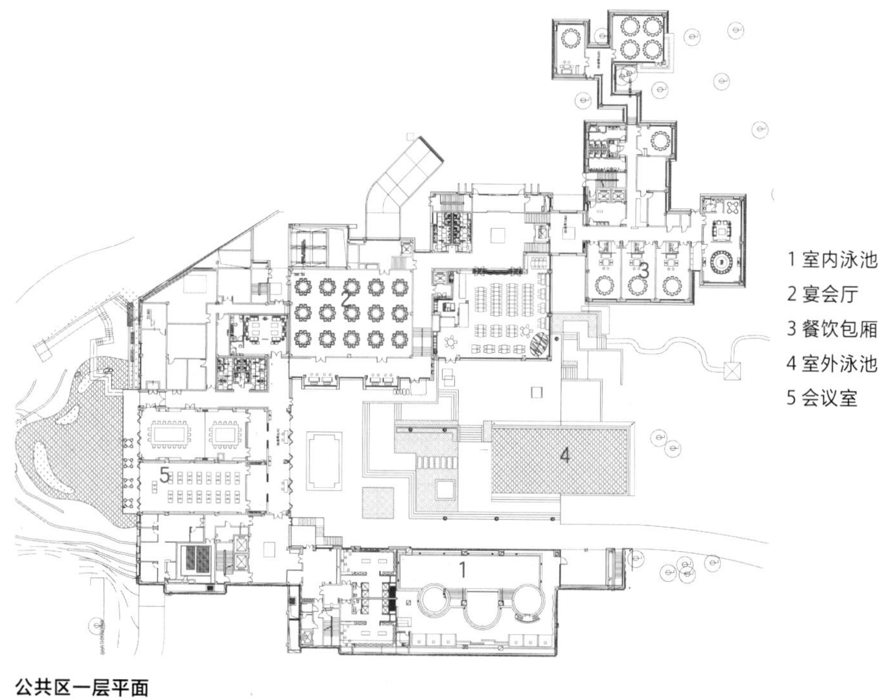

公共区一层平面

设计尊重场地山谷地貌，尽可能保留场地内生长状况良好的高大乔木、原生植被和现有的溪流，保护原有自然生境，延续环境中的原生地景。其次从水平向（由外而内）与垂直向（从下到上）抽取归纳不同层级的空间，建筑体验将在山景与湖景之间，沿着丰富的空间维度展开，并随之修正空间朝向，回避不利景观因素，最大化发挥场地的景观优势。

The design respects the valley topography of the site, preserving as much as possible the tall trees, native vegetation and existing streams that grow well on the site, protecting the original natural habitat and continuing the native landscape in the environment. Secondly, by extracting and summarizing different levels of space horizontally (from the outside to the inside) and vertically (from the bottom to the top), the architectural will be unfolded along the rich spatial dimensions between the mountain view and the lake view, and the spatial orientation will be revised accordingly, avoiding the unfavorable landscape factors, and maximizing the advantages of the site's landscape.

斜塘老街四期

XIETANG OLD STREET PHASE Ⅳ

项目类型：商业街区　　**项目地点：**江苏苏州
用地面积：50157m²　　**建筑面积：**27768m²
设计时间：2012年　　**竣工时间：**2014年

2019年度全国优秀工程勘察设计传统建筑三等奖
2020年度江苏省第十九届优秀工程设计三等奖
2020年度江苏省优秀工程勘察设计传统建筑一等奖

斜塘老街充分利用规划区别致的自然景观，挖掘悠久的历史文化积淀，传承历史文脉，建成了以高档休闲、娱乐、旅游功能为主，以传统水乡风貌为特色的商业街区。整个街道呈“河路并行”的格局，北侧内部车道沿线是餐饮区，交通便利。往南过桥是购物区，购物区排布紧凑，街区尺度宜人。与商业区背弄相隔是酒店区，酒店区与餐饮区隔河相望，酒店区主入口位于两桥之间，连接斜塘河南岸，酒店区注重内部环境的营造。酒店区以东是高端会所，三面临水，环境优美。

Xietang Old Street has made full use of the natural landscape of planning distinction, explored the long historical and cultural accumulation, inherited the historical culture, and built a commercial area with high-quality leisure, entertainment and tourism functions, featuring the traditional water city style. The whole street is in the pattern of "parallel river and road", and the restaurant area is along the inner lane on the north side, with convenient traffic. On the south side across the bridge is the shopping area, which is compactly arranged and has a pleasant scale. Separated from the commercial area is the hotel area. The hotel area is located across the river from the restaurant area. The main entrance of the hotel area is located between the two bridges, connecting the south bank of Xietang River, and the hotel area focuses on creating the internal environment. To the east of the hotel area is the high-end clubhouse, which faces the water on three sides and has beautiful scenery.

区位

设计注重生活体验和人文感受。漫步在老街上，穿梭于传统与现代之间，老街与周边环境自然融合，演绎着一幕幕生动的生活情景，这就是斜塘老街的意义所在。

The design focuses on life experience and humanistic feeling. Walking along the old street, shuttling between tradition and modernity, the natural integration of the old street and the surrounding environment interpret a vivid scene of life. That is the meaning of Xietang Old Street.

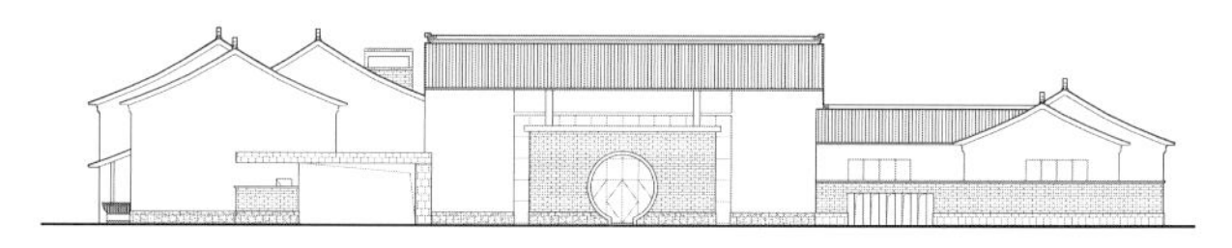

酒店东立面

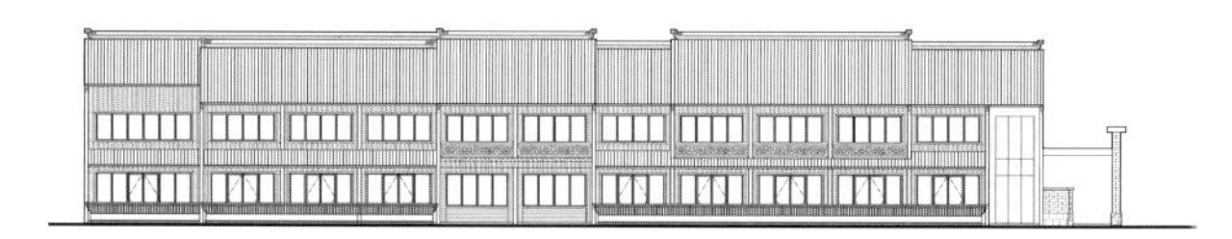

酒店南立面

酒店北立面

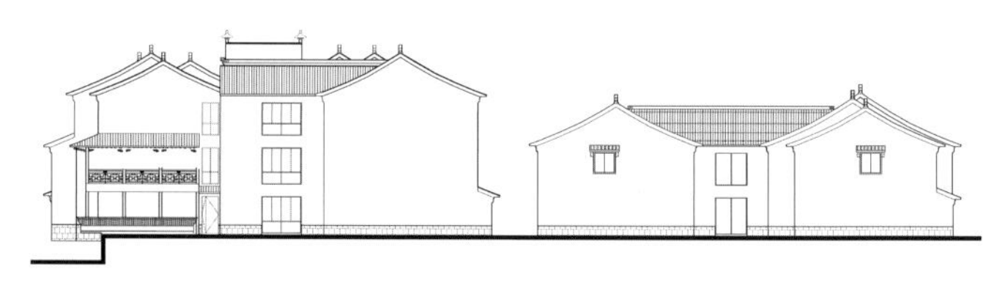

酒店西立面

1 大堂　　　4 客房区
2 早餐厅　　5 天井
3 西餐备餐间

酒店一层平面

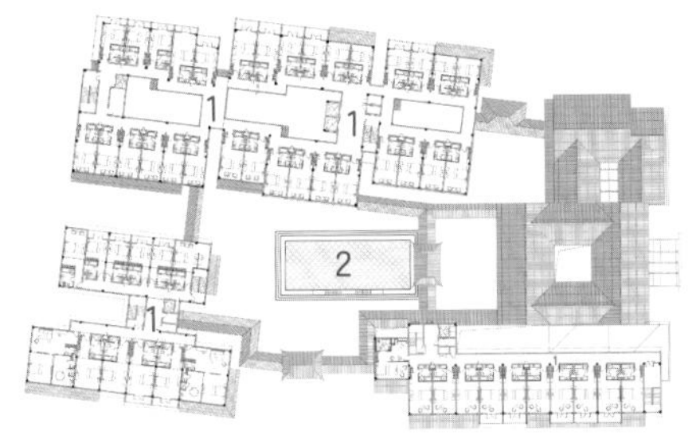

1 客房区　　2 室外游泳池

酒店二层平面

雅致·湖沁阁酒店坐落在街区中央，酒店建筑布局采用傍水造园手法，分散围合设置，内部以游廊的形式串联，配以景观设计，达到传统园林的视觉效果。一步一景，移步异景，苏州园林的精华便浓缩在这小小的院子中。借鉴中国园林艺术的现代酒店建筑，探索园林艺术和建筑空间艺术的内涵，巧妙地运用层次变化规律，使有限空间小中见大，局促中见舒展，营造出一种深邃幽远的氛围。

Mehood Elegant Hotel is located in the middle of the street, the hotel building layout adopts the waterfront gardening skills to set up dispersed enclosure. The interior is arranged in the form of corridors with landscaping to achieve the visual effect of traditional gardens. One step, one view. The essence of Suzhou gardens is concentrated in this small courtyard. The modern hotel architecture draws on Chinese garden art to explore the connotation of garden art and architectural space art. The limited space can be seen bigger picture in small matters, which create a deep and distant atmosphere through the skillful use of the hierarchical change law.

无锡尚德太阳能电力研发办公楼

WUXI SUNTECH R&D OFFICE BUILDING

项目类型：科技研发　　**项目地点**：江苏无锡

用地面积：44510m²　　**建筑面积**：74680m²

设计时间：2006年　　**竣工时间**：2009年

合作单位：（奥地利）Wolfgang Wieser Mario Buchegger J. W. Tak

2011年度中国建筑学会优秀建筑结构设计三等奖

2011年度中国建筑学会建筑创作佳作奖

2010年度江苏省第十四届优秀工程设计一等奖

2009年度江苏省优秀勘察设行业奖结构一等奖

2010年度江苏省城乡建设系统优秀勘察设计一等奖

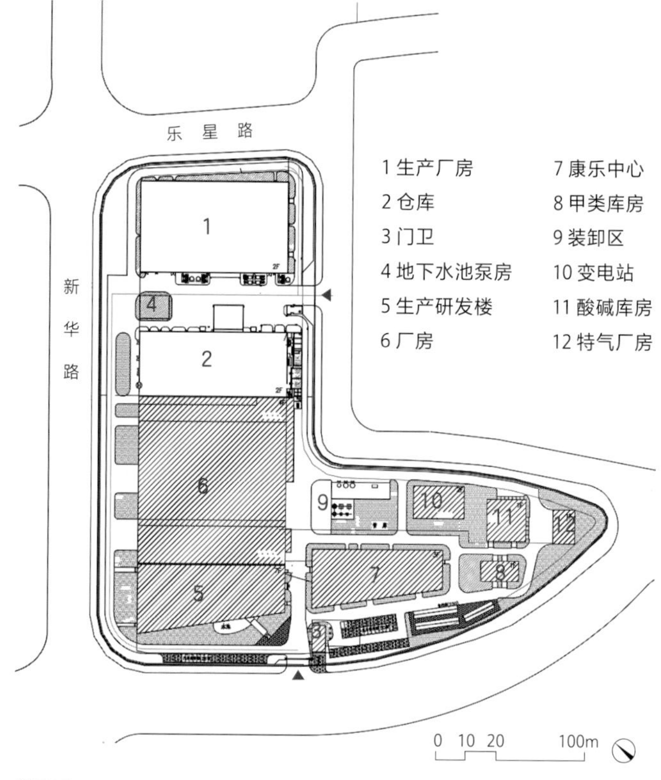

总平面

项目位于无锡高新技术产业开发区新华路，为体现企业的特色技术和对可再生能源利用的长远愿景，大楼设计被赋予了"零能耗"、"功能型"、"生态建筑"的概念，成为集建筑形态空间审美与光伏应用示范、光伏技术研发于一体的综合性绿色建筑示范项目。

建筑内部空间将每个楼层的各区域互相错动咬合，形成一个具有丰富空间感的高大空间，如同在大厅内飘浮或停靠在岸边的一艘艘小船。太阳能光电板取代传统玻璃幕墙，作为主立面外墙围护结构，太阳能电池板墙以一种简洁有力的形象拔地而起，极富视觉冲击力。在体现建筑的现代感和高科技含量的同时，作为研发办公楼的能源站，其照明、热水等能量都来源于取之不尽、用之不竭的太阳能。

The project is located in Xinhua Road, Wuxi Hi-Tech Industrial Development Zone. In order to reflect the company's characteristic technology and long-term vision of renewable energy utilization, the building design is endowed with the concepts of "Zero Energy Consumption", "Functional" and "Eco-Building", and it becomes a comprehensive green building demonstration project integrating the spatial aesthetics of architectural form, demonstration of photovoltaic application, and R&D of photovoltaic technology.

The internal space of the building is a staggered combination of various areas on each floor, forming a tall space with a rich sense of space on a single level, just like boats floating in the hall and resting on the shore. Solar photovoltaic panels to replace the traditional glass curtain wall, as the main facade wall enclosure, solar panel wall with a simple and powerful image rises up, very rich in visual impact. While reflecting the modernity and high-tech content of the building, as the energy station of the R&D office building, its lighting, hot water and other energy come from the inexhaustible solar energy.

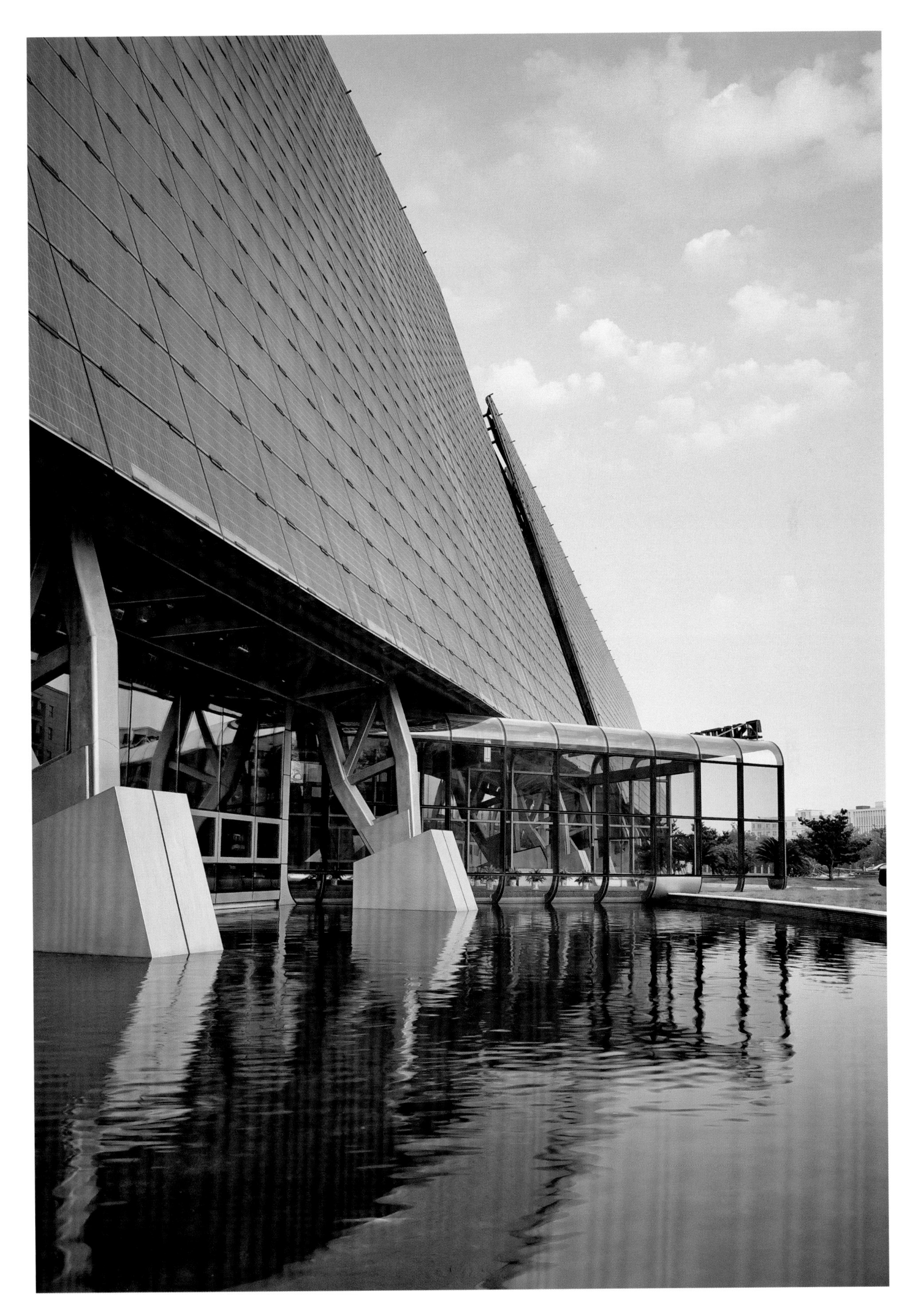

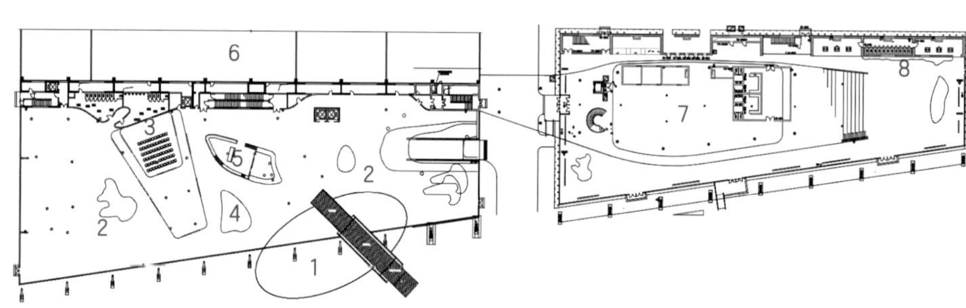

1 水面
2 产品展览大厅
3 临时产品展示影院
4 接待
5 接待办公室
6 生产厂房
7 餐厅
8 攀岩

一层平面

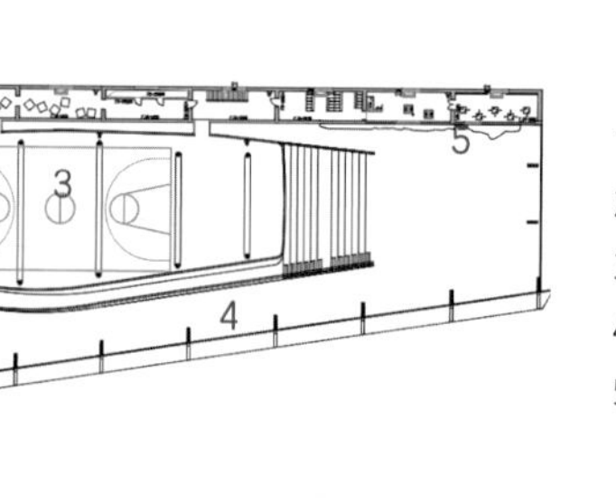

1 研发办公室
2 培训室 / 会议室
3 篮球场
4 共享空间
5 攀岩

二层平面

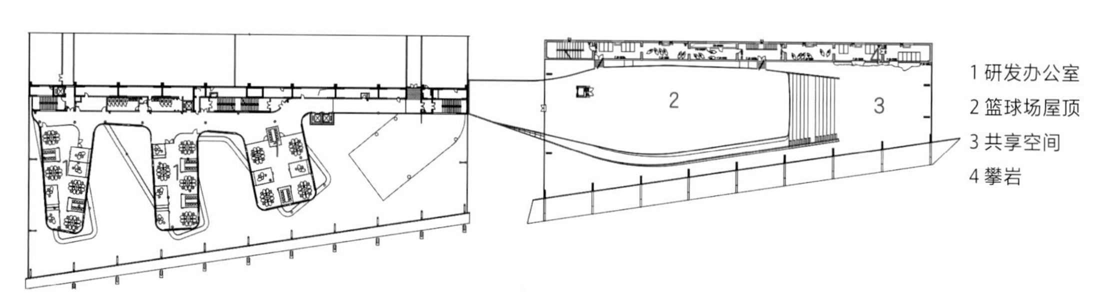

1 研发办公室
2 篮球场屋顶
3 共享空间
4 攀岩

三层平面

结合建筑独特的内部空间设计，结构设计采用可重复循环利用的全钢结构体系，满足建筑大跨度、大空间、大悬挑的空间效果要求，使建筑材料的特性得到最大发挥。在机电设备系统方面，采用多种措施综合运用，例如地源热泵系统和空调全热回收系统的运用等，在太阳能利用的基础上，进一步推进了可再生能源利用在本建筑中的实践，使研发办公楼最终实现“零能耗”目标。

Combined with the building's unique interior space design, the structural design adopts an all-steel structural system that can be reused and recycled. The design meets the requirements of large span, large space, large overhanging space effect of the building, so that the characteristics of the building materials are maximized. In terms of electromechanical equipment systems, a variety of measures are used, such as the use of ground-source heat pump systems and air-conditioning heat recovery systems, etc., and on the basis of solar energy utilization, the use of renewable energy in the building is further promoted, so that the R&D office building ultimately realizes the goal of "zero energy consumption".

黄山小罐茶运营总部

HUANGSHAN XIAO GUAN CHA OPERATION HEADQUARTERS

项目类型：研发中心/产业园　　**项目地点**：安徽黄山

用地面积：213333m²　　**建筑面积**：140000m²

设计时间：2017年　　**竣工时间**：2022年

2022年度江苏省土木建筑学会第十六届“建筑创作奖”一等奖

项目位于黄山高铁站进市区沿线，是黄山经济开发区的门户，项目的整体风貌对于经济开发区形象的塑造至关重要。如何在满足工厂高效生产需求的同时，展示当代智慧工厂先进的生产力并反映品牌的精神气质，成为本项目最大的挑战。设计在总体规划上摒弃传统工厂的常规布局，通过对工艺流线、人员动线、货运流线的重新梳理，创新地将生产线更安全友好地呈现给公众和城市，并运用巧妙的设计手法赋予工厂新的价值和形象，将厂区打造成展现公司品牌的新场所、代表企业实力的开发区新名片和激发都市文化的新城市空间。

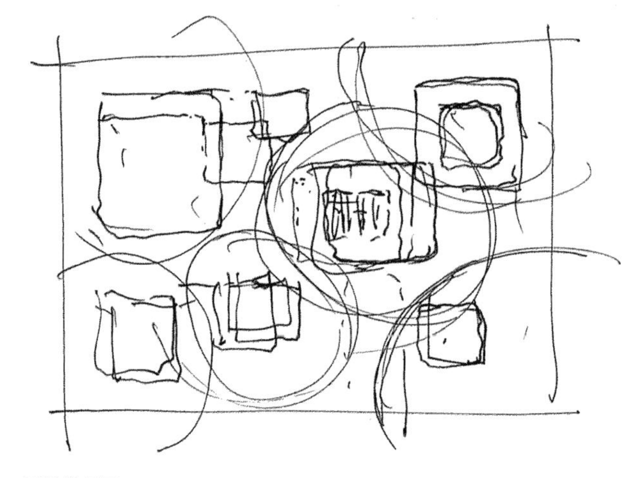

手绘总平面

The project base is located along the railroad line in Huangshan and it is the gateway to the Huangshan Economic Development Zone. The overall appearance of the project is critical to the image of the Economic Development Zone. How to meet the factory's efficient production needs while demonstrating the advanced productivity of a contemporary smart factory and reflecting the spirit of the brand became the biggest challenge of this project. The design abandons the conventional layout of traditional factories in the overall planning, and innovatively presents the production line to the public and the city in a safer and friendlier way through the reorganization of process flow, personnel flow, and freight flow. It also uses skillful design techniques to give the factory a new value and image, turning the factory into a new place to show the company's brand, a new business card of the development zone representing the strength of the company, and a new urban space to stimulate the urban culture.

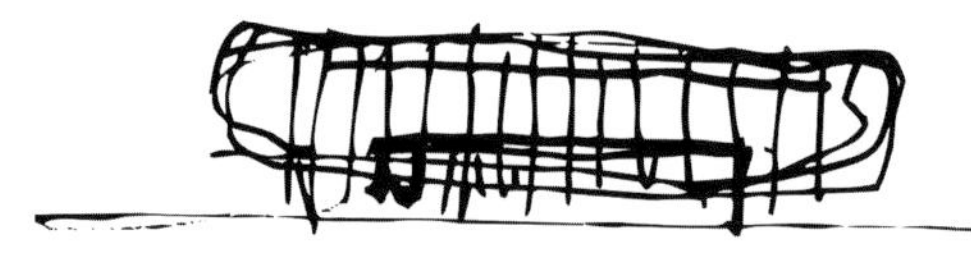

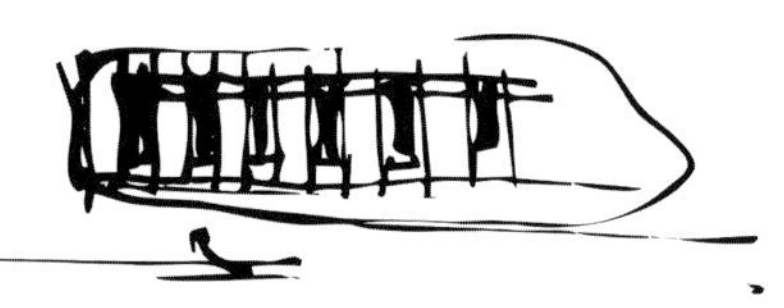

手绘立面

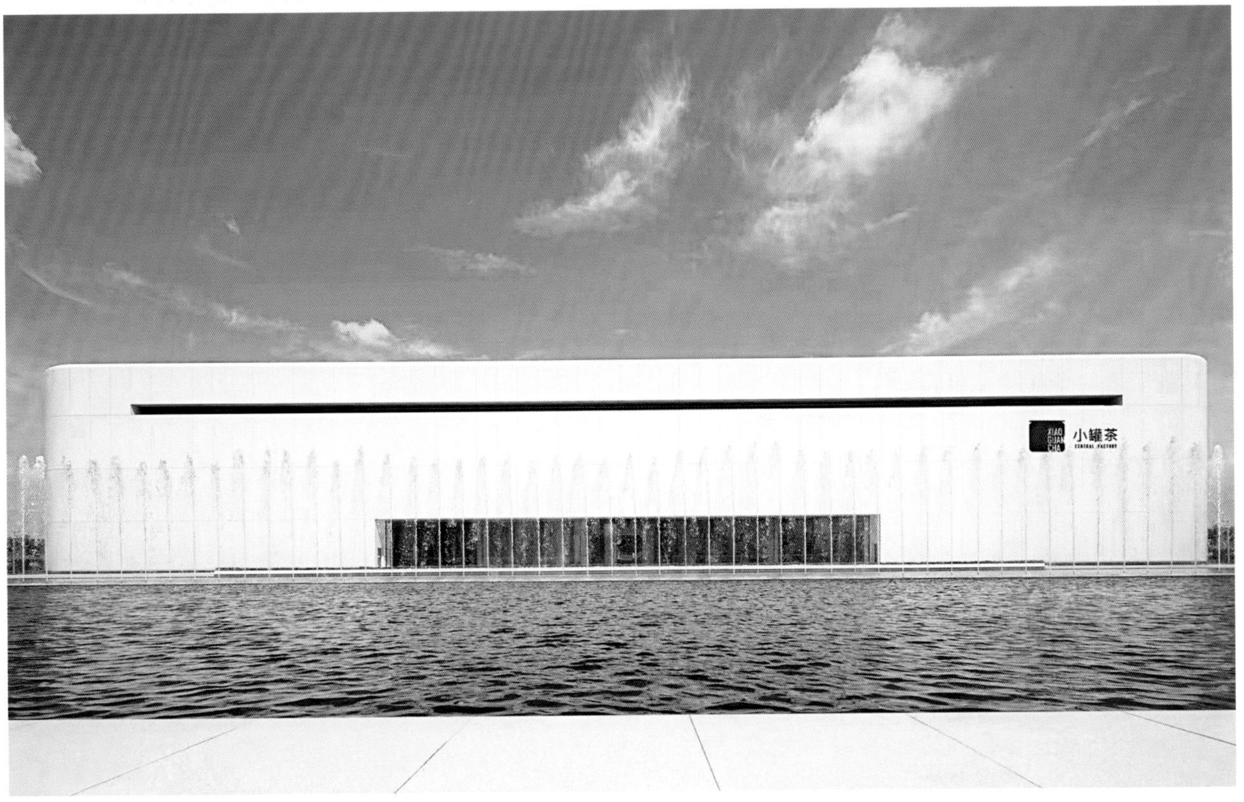
XIAO
GUAN
CHA
小罐茶
CENTRAL FACTORY

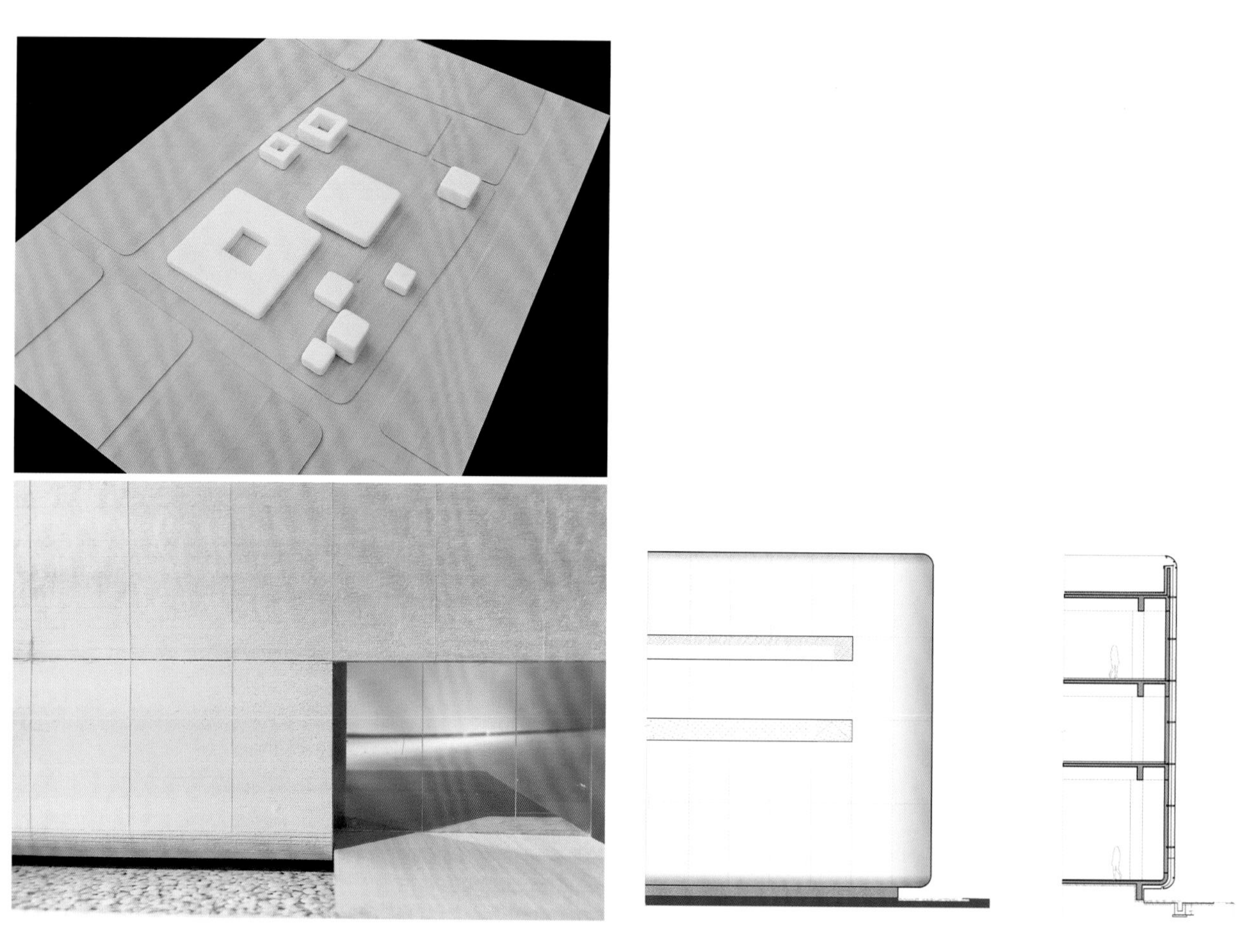

模型照片

1# 立剖面局部大样

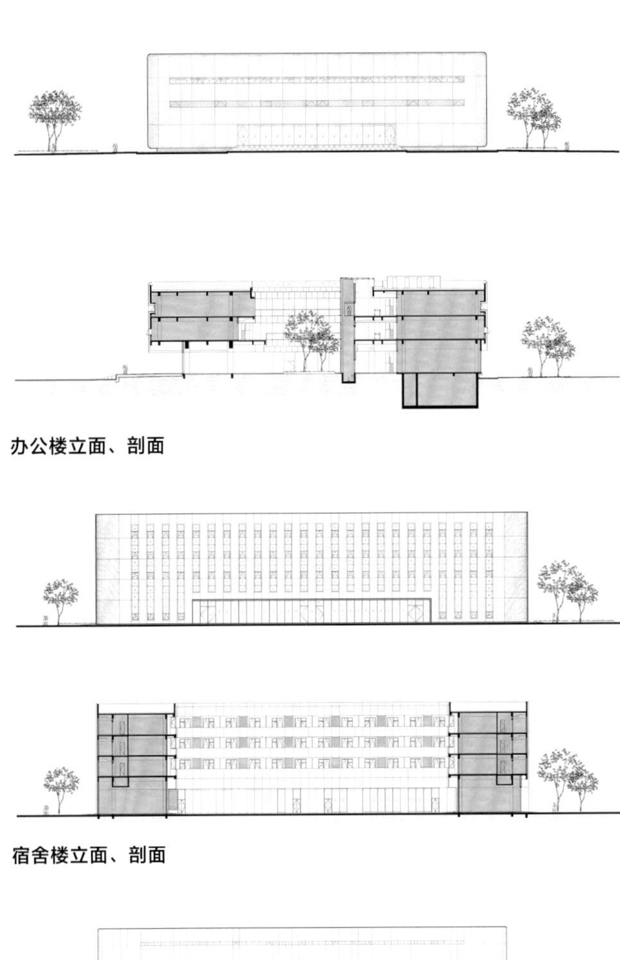

办公楼立面、剖面

宿舍楼立面、剖面

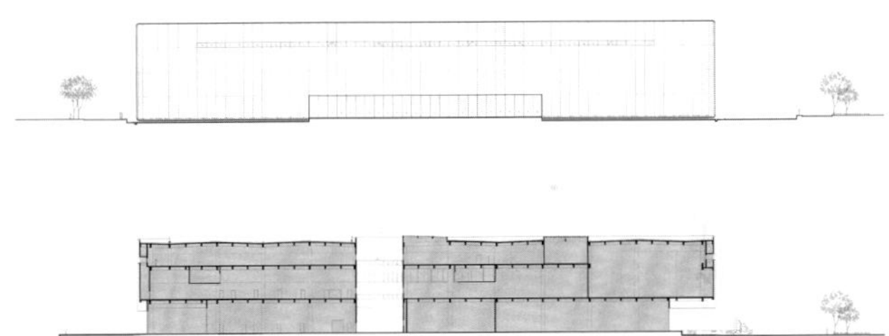

工厂仓库立面、剖面

设计从传统徽派村落汲取灵感，将建筑组群有机地散落布局，高低错落，与周边山体更加融合。办公楼设计运用了中央庭院和室外环廊，模糊了室内外分界，将自然引入空间。建筑设计在最简单高效的方形基础上反复推敲，最终用圆角取代了直角，轻巧的弧形处理使厚重的工厂呈现一种漂浮的效果，也更符合小罐茶精致温润的产品精神。立面材料采用大型纯白色蜂窝铝板，使建筑整体形象更加纯粹简约。

The design draws inspiration from the traditional Hui-style villages, and the building groups are organically dispersed in a layout with staggered heights to blend more with the surrounding mountains. The design of the office building utilizes a central courtyard and an outdoor wrap-around corridor, blurring the distinction between indoors and outdoors and bringing nature into the space. Based on the traditional square repeatedly extrapolated the subtle design changes, and finally replaced the square right angle with a circular guide angle. The light curved shape presents a floating effect to a very heavy factory, so that the building is simple and delicate which correspond to the delicate and gentle product connotation of Xiaoguancha products. Large and white honeycomb aluminum panels were used for the facade material, making the overall image of the building more pure and simple.

冯梦龙村全域旅游发展规划

FENGMENGLONG VILLAGE OVERALL TOURISM DEVELOPMENT PLAN

项目类型：乡村振兴　　**项目地点：**江苏苏州

用地面积：3800000m^2　　**竣工时间：**2019年

设计时间：2018年

冯梦龙村自然和人文底蕴深厚，规划以乡村文旅为主线，发展生态休闲农业，打造一个集冯梦龙精神传承、生态休闲、亲子游乐于一体的景区。根据“一核、一环、两轴、三片区”的村域空间结构规划，对道路系统、旅游线路、房屋整体风貌、小道景观、村庄设施等多方面、多维度进行整体综合提升，打造美丽宜居的特色田园乡村，开创农业、文化、旅游三方融合的乡村全域发展新模式。

Fengmenglong Village has a deep natural and cultural heritage, and it is planned to develop ecological leisure agriculture with rural cultural tourism as the main line, to create a scenic spot that integrates Fengmenglong's spiritual heritage, ecological leisure and parent-child fun. According to the spatial structure planning of "one core, one ring, two axes, three areas", the road system, tourism route, overall house style, trail landscape, village facilities and other aspects, multi-dimensional to carry out overall comprehensive improvement. It is committed to creating a beautiful and livable rural village with special characteristics, and creating a new model of whole-area development of the countryside with tripartite integration of agriculture, culture and tourism.

规划总平面

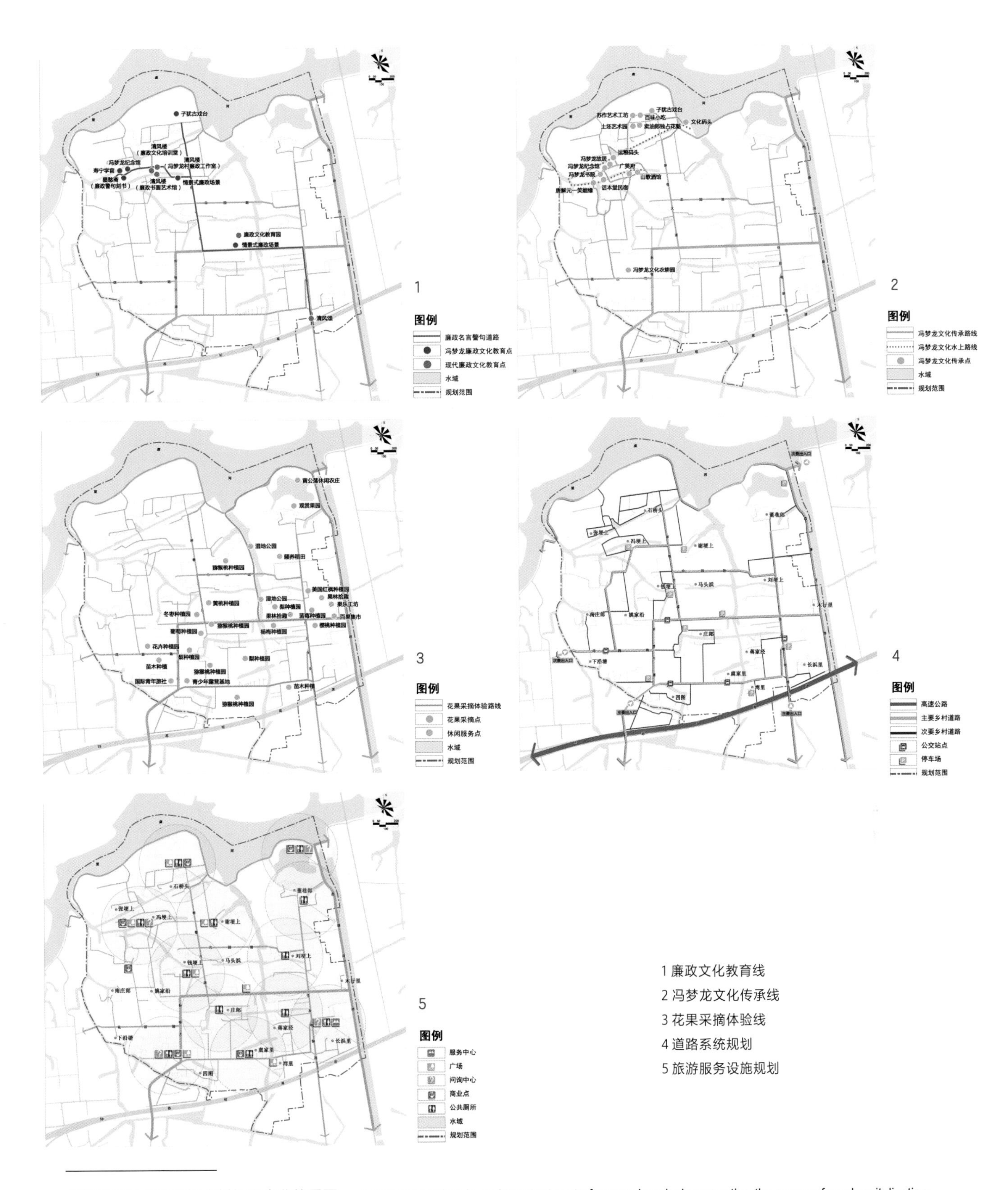

1 廉政文化教育线
2 冯梦龙文化传承线
3 花果采摘体验线
4 道路系统规划
5 旅游服务设施规划

全域旅游是全面推进乡村振兴事业的重要组成部分，也是实现乡村振兴的主要动力和保障。在乡村振兴的发展背景下，冯梦龙村紧抓特色田园乡村示范点和乡村全域融合发展的机遇，深入挖掘整合其特有的名人文化和产业资源，通过依托交通区位的便利优势，借力“冯梦龙”历史人文效应，做足乡村旅游文章。

Territorial tourism is an important part of comprehensively promoting the cause of rural revitalization, and it is also the main driving force and guarantee to realize rural revitalization. Under the development background of rural revitalization, Fengmenglong Village grasps the opportunity of the demonstration site of distinctive rural village and the integration development of the whole rural area, and deeply explores and integrates its unique celebrity culture and industrial resources. By relying on the advantages of convenient transportation location, rural tourism articles can take advantage of the "Fengmenglong" to do a full historical and humanistic effect.

设计团队从一开始就制定了规划、建筑、景观联合设计方针，旨在打造可以落地的方案。团队首先完成冯埂上特色田园乡村规划，归纳总结村庄特色，提炼价值爆点。随后开展全域旅游发展规划，规划落地后，进行深入的建筑、景观方案深化与施工图设计。目前团队已完成冯梦龙村全域农文旅发展规划、冯埂上特色田园乡村、冯梦龙纪念馆、冯梦龙山文化歌馆、卖油郎油坊等项目，助力冯梦龙村走出一条独具特色的乡村振兴之路。

From the beginning, the design team developed a common design policy of planning, architecture and landscape to create a solution that could be landed. The team first completed the planning of the special rural village of Fenggengshang, summarizing the village features and refining the value bursts. The team then carried out the planning of the whole area tourism development, and after the planning landed, conducted in-depth architectural and landscape plan and construction drawing design. At present, the team has completed the whole area agricultural, cultural and tourism development planning of Fengmenglong Village, Fenggengshang Featured Countryside, Fengmenglong Memorial Hall, Fengmenglong Mountain Cultural Song House and Oil Selling Workshop, helping Fengmenglong Village to come out of a unique road of rural revitalization.

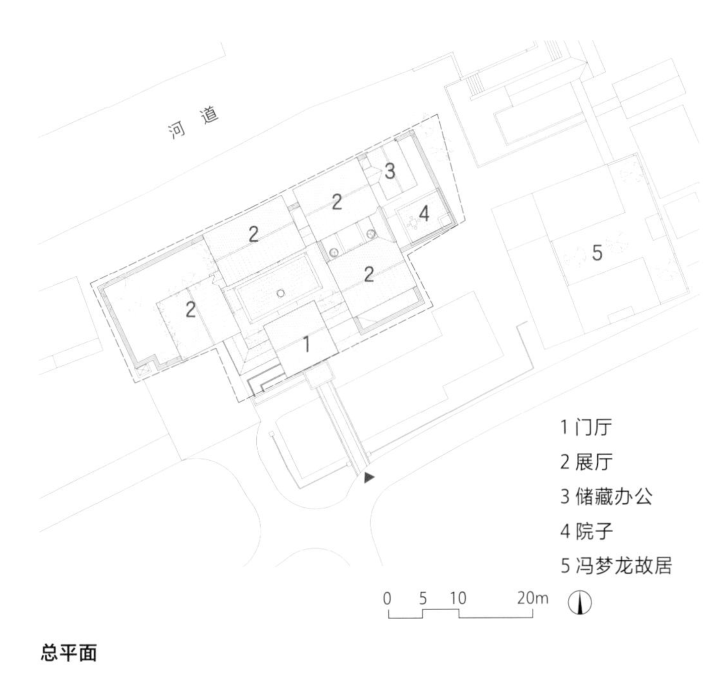

总平面

冯梦龙纪念馆位于苏州相城区黄埭冯梦龙村冯埂上，设计以传统苏州民居为原点，考量了适应环境及烘托人物性格两个因素，放弃追求标新立异而选择小巧质朴，保证游览的连续性及完整性。设计营造古朴淡雅的建筑风格，体现淡泊而亲和的中国传统文人哲学，借以纪念冯梦龙“存仁积善”“以廉代匮”的高尚品格，彰显冯梦龙其人其品以及对其产生深远影响的吴地文化。

2019年度全国优秀工程勘察设计行业奖传统建筑二等奖
2019年度江苏省城乡建设系统优秀勘察设计奖一等奖
2020年度江苏省第十九届优秀工程设计一等奖

Fengmenglong Memorial Hall is located in Fenggengshang, Fengmenglong Village, Huangdai, Xiangcheng District, Suzhou. The building takes the traditional Suzhou residence as the design origin and considers the two elements of adapting to the environment and bringing out the character of the person. The new and unusual is abandoned in favor of the small and simple, the continuity and integrity of the tour is maintained, and the attempt is made to create the atmosphere that the traditional architecture can express. It reflects the traditional Chinese literati philosophy of indifference and affability, and commemorates Fengmenglong's noble character of "Saving benevolence and accumulating goodness" and "Making up for local poverty with the integrity of officials".

党建文化馆二期位于冯梦龙纪念馆西南侧，设计将古法榨油与冯梦龙故事相结合，为空置的老旧民居植入“油坊”这一新型功能业态，使其成为“农、文、旅”三要素有机结合的空间载体。设计通过对场地的调研以及对空间形态的分析，保留北侧两栋一层民房，对南侧两层民房进行重建，采用低技、乡土的建造手段，不仅维持了原有场地的场所感，保留了原有场地的空间记忆，也体现了村舍的本真面貌。

Phase II of the Party Building Cultural Center is located in the southwest of the Fengmenglong Memorial Hall. The design combines the ancient method of oil extraction with the story of Fengmenglong, and the vacant old residential houses are implanted the "oil mill" as a new functional business, making it a space carrier that organically combines the three elements of "agriculture, culture and tourism". Through researching the site and analyzing the spatial patterns, the design retained the two one-story houses on the north side, the two-story residential house on the south side was rebuilt. By adopting the lowest grade of the ancient construction system, it not only maintains the sense of the original site and preserves the spatial memory, but also reflects the true face of the houses.

1 党建文化馆　2 磨粉　3 菜地

总平面

2020年度江苏省第十九届优秀工程设计二等奖

2020年度江苏省城乡建设系统优秀勘察设计奖二等奖

山歌文化馆位于冯埂上村口，是冯埂上的展示窗口以及村民的公共活动空间。设计根据功能空间将四个组团由西向东一字排开，在组团体量之间镶嵌透明的玻璃盒子，形成对现代建筑语言的演绎。设计采取保持原肌理修复的方法，充分体现低技环保的地域建筑特色，尤其对传统砖砌技术进行研究和利用。广笑府南侧以整面清水镂空砖墙作为主立面，构成建筑与南侧田野之间的对话，也使得广笑府成为冯埂上的标志性建筑。

The Folk Song Culture Hall is located at the entrance of Fenggengshang Village, which will become the window of Fenggengshang and a public activity space for "villagers". According to functional spaces the four clusters were lined up from west to east with transparent glass boxes among the group volumes, and made up the interpretation of modern architectural language at the design level. The project fully reflects the characteristics of low-tech and eco-friendly regional architecture, especially the study and use of traditional masonry techniques. The entire plain hollow brick wall is used as the main facade on the south side of Guangxiaofu, creating a dialog between the building and the fields on the south side, and also making Guangxiaofu a landmark building on Fenggengshang.

2023年度江苏省第二十届优秀工程设计村镇二等奖
2021年度江苏省城乡建设系统优秀勘察设计村镇二等奖
2021年度江苏省优秀工程勘察设计行业奖传统建筑二等奖
2020年度江苏省土木建筑学会第十四届“建筑创作奖”乡村建筑一等奖

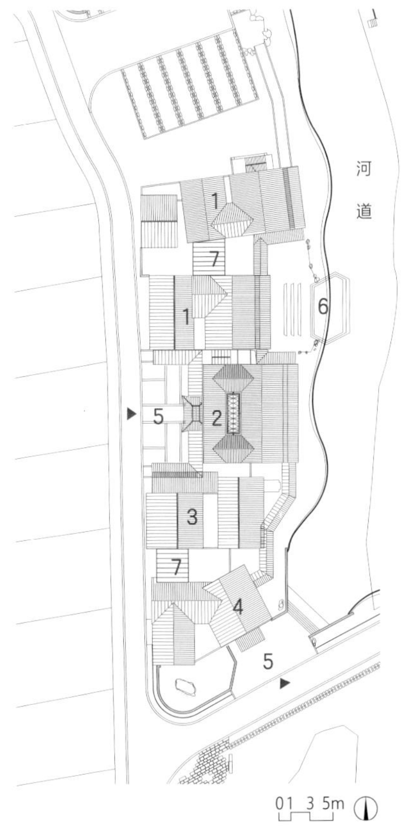

1 山歌馆
2 广笑府
3 多功能活动中心
4 游客中心
5 广场
6 戏台
7 玻璃盒子

总平面

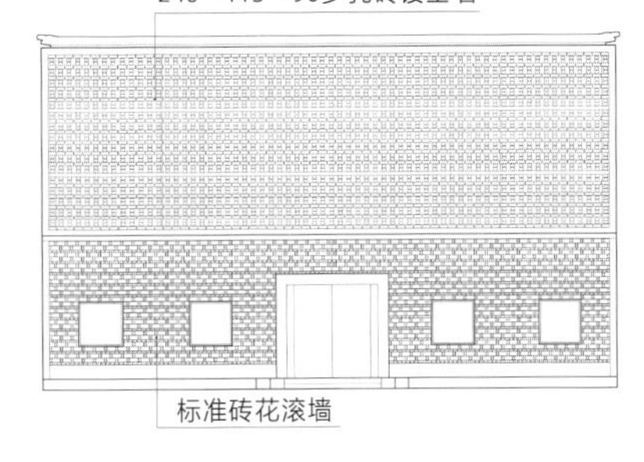

砖墙立面示意

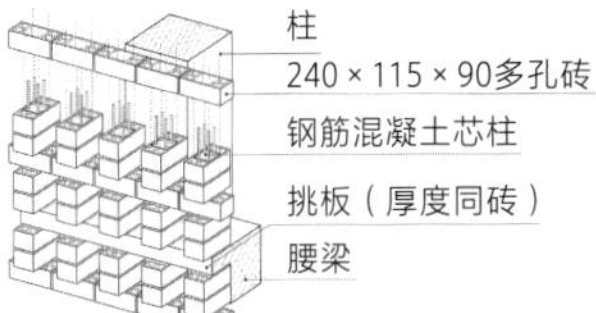

砌砖示意

昆山杜克大学二期

DUKE KUNSHAN UNIVERSITY PHASE Ⅱ

项目类型：教育建筑　　**项目地点**：江苏苏州
用地面积：189334m²　　**建筑面积**：95007m²
设计时间：2019年　　**竣工时间**：2023年
合作单位：帕金斯威尔建筑设计咨询（上海）有限公司

2022年度第九届江苏省勘察设计行业信息模型（BIM）应用大赛一等奖

项目位于江苏省昆山市杜克大道北侧。项目致力打造和谐美好的校园环境，同时在已建成的一期校园和计划建设的三期校园之间起到承上启下的作用。

The project is located on the north side of Duke Avenue in Kunshan City, Jiangsu Province. The project aims to create a harmonious and beautiful campus environment while serving as a continuum between the completed Phase Ⅰ campus and the planned Phase Ⅲ campus.

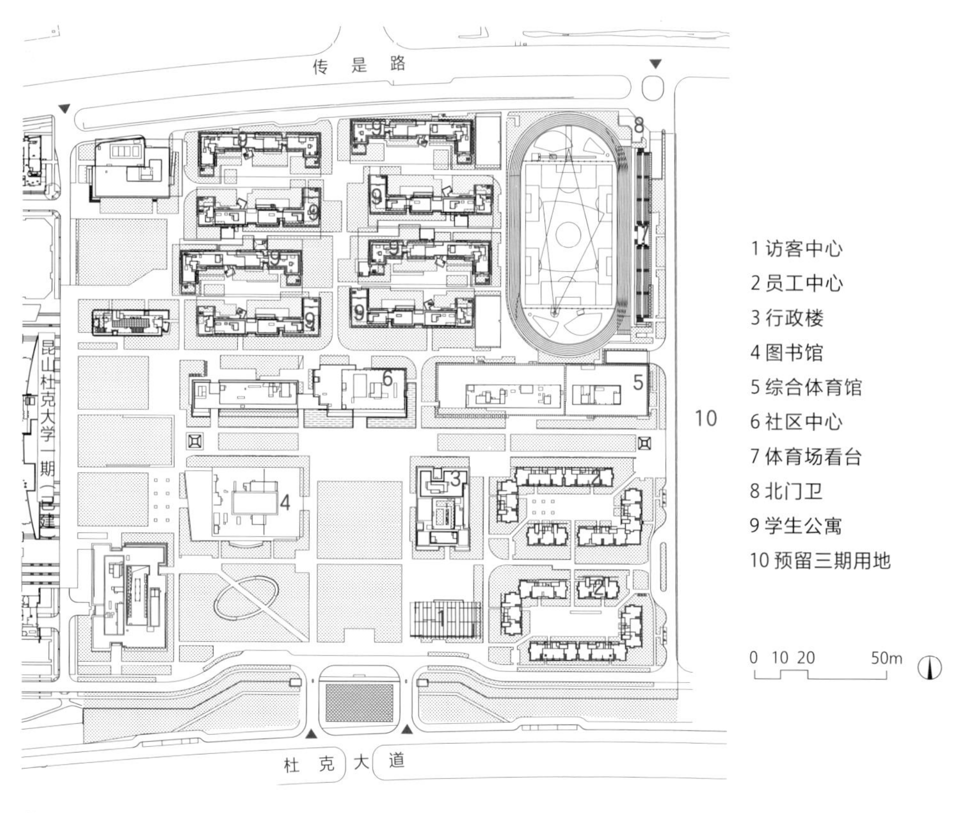

总平面

建筑均为多层建筑，外形体量富有雕塑感，形成丰富的光影效果。主要教学建筑的材质和色彩与一期的建筑风格呼应，立面材质为石材、金属和玻璃。平面的模数化设计赋予使用功能的灵活性和未来空间调整的弹性，各单体设计强调室内主空间与室外景观环境的关联与对话。通过节约能源、降低能耗的多种设计手法，营造出优雅和实用兼备的现代化校园环境。

The buildings are all multi-story, with sculptural volumes that create rich light and shadow effects. The materials and colors of the main teaching building echo the architectural style of the Phase 1, with facade materials of stone, metal and glass. The modal design of the plane gives flexibility of use and future space adjustment. The design emphasizes the connection and dialogue between the main indoor space and the outdoor landscape environment. Through a variety of design techniques to conserve energy and reduce energy consumption, the designers have created a modern campus environment that is both elegant and functional.

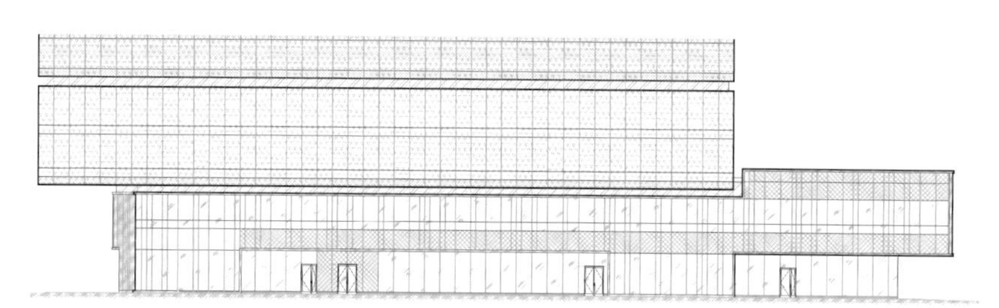

图书馆 A ~ F 轴立面

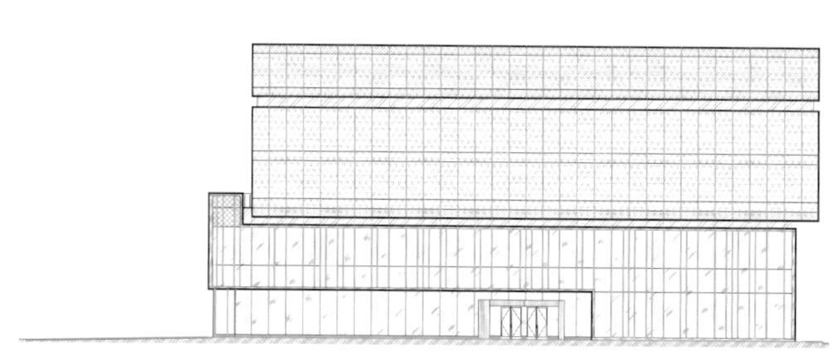

图书馆 9 ~ 1 轴立面

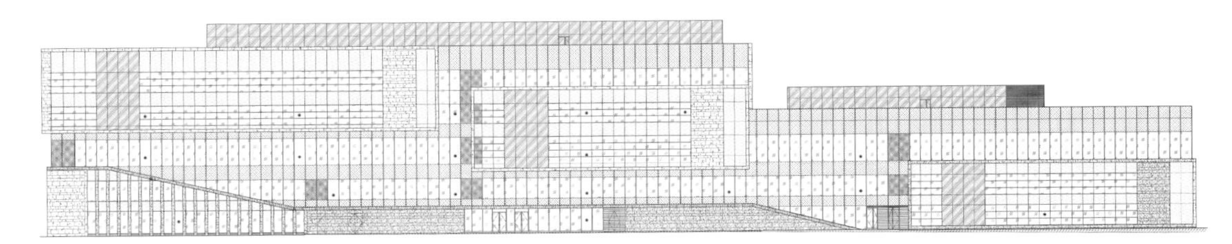

综合体育馆南立面

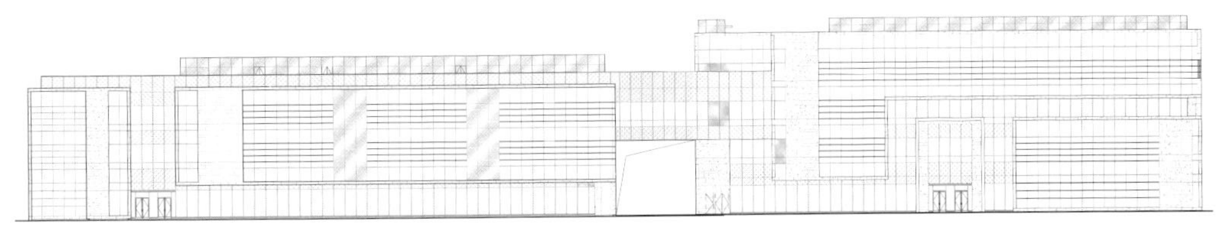

社区中心立面

中国中医科学院西苑医院苏州医院

SUZHOU HOSPITAL OF XIYUAN HOSPITAL OF CHINA ACADEMY OF CHINESE MEDICAL SCIENCES

项目类型：医疗建筑　　**项目地点：**江苏苏州

用地面积：56643m²　　**建筑面积：**206000m²

设计时间：2020年　　**竣工时间：**在建

项目为苏州市与中国中医科学院西苑医院合作创建的区域中医诊疗中心，力求打造“立足苏南、辐射长三角、影响全国、面向国际”的一流国家区域医疗中心样板医院。

项目难点在于要跨越南北两个地块将复杂的医疗功能与流线进行合理规划与组织，并对场地内进行既有地下桩基工程的最大化利用。设计按医疗功能使用需求及未来发展进行组合布局，前低后高，形成良好的空间关系和层次。两个地块之间由连廊、路面桥梁及地下通廊连接。

The project is a regional TCM diagnostic and treatment center created in cooperation between Suzhou and Xiyuan Hospital of China Academy of Chinese Medical Sciences, aiming to build a first-class national regional medical center model hospital "based in Southern Jiangsu, radiating the Yangtze River Delta, influencing the whole country, and facing the international world".

The challenge of the project was to rationally plan and organize the complex medical functions and flow lines across the north and south parcels, and to maximize the use of the existing underground piling works on the site. The design combines the layout according to the needs of medical function use and future development, with low front and high back, forming a good spatial relationship and hierarchy. The two parcels are connected by connecting corridors, roadway bridges and underground breezeways.

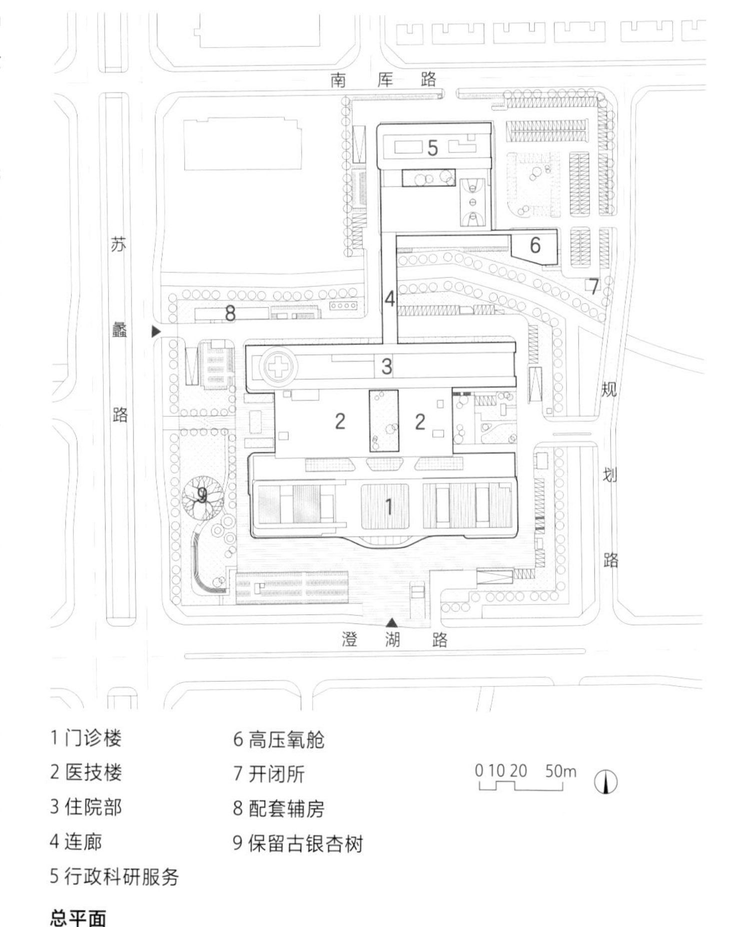

1 门诊楼
2 医技楼
3 住院部
4 连廊
5 行政科研服务
6 高压氧舱
7 开闭所
8 配套辅房
9 保留古银杏树

总平面

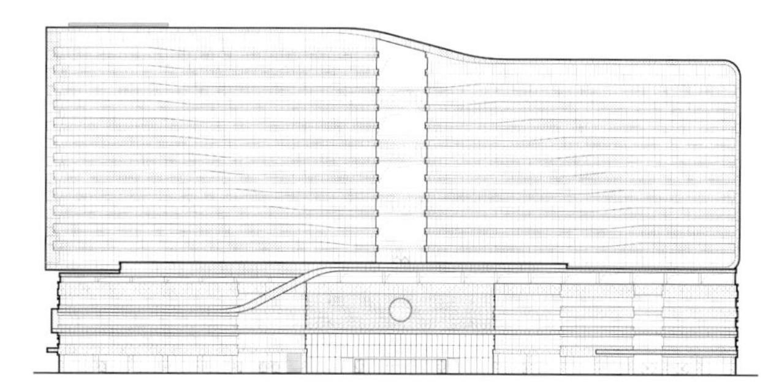

医疗综合楼北立面

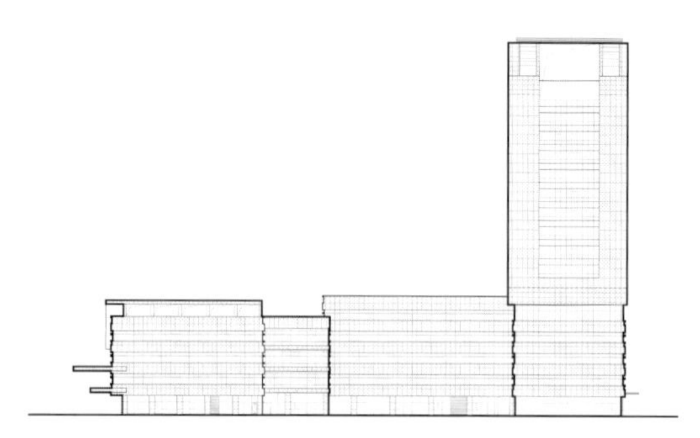

医疗综合楼东立面

在项目总体布局上，功能分区明确，流线组织合理，空间环境舒适，统筹考虑建筑、场地、景观绿化之间的和谐关系。在平面布局上，强调分区明确、医患分流、标准设计、灵活多变的设计理念。在空间环境上，体现“以人为本”的思想，努力为患者创造优美舒适的就医体验，并为医护及管理人员提供高效安全的工作环境，充分利用集中绿地、屋顶花园、下沉庭院、公共大厅等空间要素，营造高效、友好、安全的医疗环境。在建筑外立面形象造型上，取意江南山水，打造契合医院人文意蕴、富有地域特色的整体建筑形象，构建集医、教、研、防、康、养于一体的医疗建筑环境。

In the overall layout of the project, the functional partition is clear, the flow is reasonably organized, the spatial environment is comfortable, and the harmonious relationship between the building, the site, and the landscape greenery is considered. In the plan layout, the design concept of clear zoning, patient-doctor triage, standard design and flexibility is emphasized. In terms of spatial environment, it reflects the idea of "people-oriented", strives to create a beautiful and comfortable medical experience for patients, and provides an efficient and safe working environment for medical and nursing staff and management personnel. The centralized green areas, roof gardens, sunken courtyards, public halls and other forms are taken full use to create an efficient, friendly and safe medical environment. In terms of the image of the building facade, it fully explores the landscape culture of Wuzhong, creates an overall architectural image that fits the humanistic meaning of the hospital and is rich in Wuzhong characteristics, and builds a medical architectural environment that integrates medicine, education, research, prevention, health and recuperation.

内
内

无锡医疗健康产业园

WUXI MEDICAL HEALTH INDUSTRIAL PARK

项目类型： 医疗建筑
用地面积： 92156m²
设计时间： 2022年
床位数量： 1500

项目地点： 江苏苏州
建筑面积： 379500m²
竣工时间： 在建
合作单位： 瑞士Lemanarc建筑设计咨询有限公司

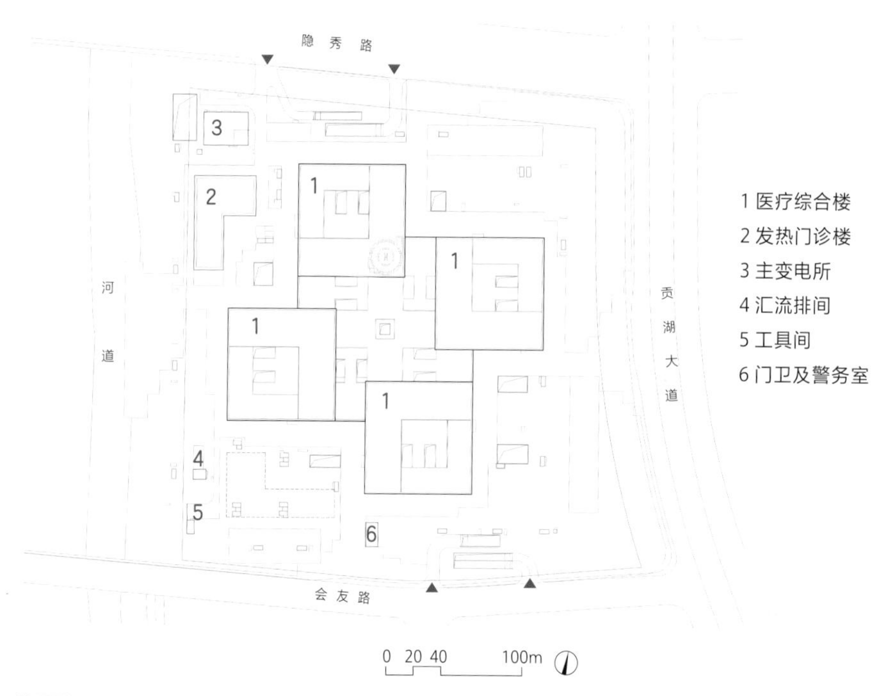

总平面

本项目旨在建成一家集妇儿预防、医疗保健、科研教学一体化发展，在长三角区域内极具影响力的三级甲等妇儿医疗保健中心。

方案以“十字风车，生生不息”为灵感，以多中心医疗园区的创新模式，巧妙结合场地与建筑功能特点，构建“四个园区、九个中心”的十字风车形布局。根据患者人群划分儿童、妇产、保健与急诊急救四大园区，多学科中心间相互独立又彼此联系，场地四周向外开放，成为一座美丽的城市客厅。

The project aims to build a Grade A tertiary maternal and child healthcare center that integrates maternal and child prevention, medical care, research and teaching, and is highly influential in the Yangtze River Delta region.

Inspired by the "cross windmill, endless life", the plan is based on the innovative model of multi-center medical park, which skillfully combines the functional characteristics of the site and buildings to build a layout of "four parks and nine centers" with a cross windmill model. According to the patient population, it is divided into four areas, namely children, maternity, health care and emergency care, and the multiple subject centers are independent and interlinked. The site opens up around the perimeter to the outside and becomes a beautiful urban living room.

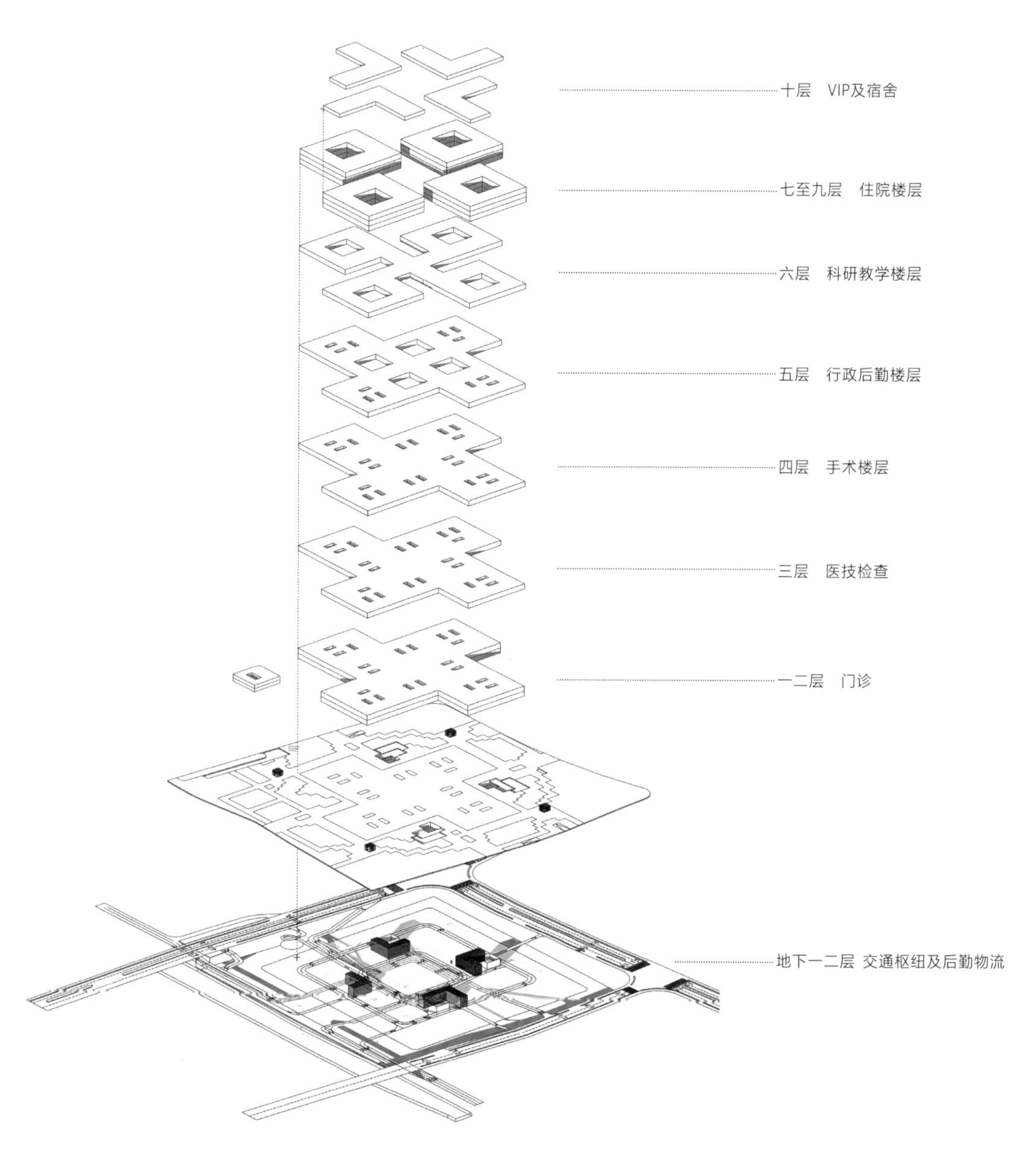
十层 VIP及宿舍
七至九层 住院楼层
六层 科研教学楼层
五层 行政后勤楼层
四层 手术楼层
三层 医技检查
一二层 门诊
地下一二层 交通枢纽及后勤物流

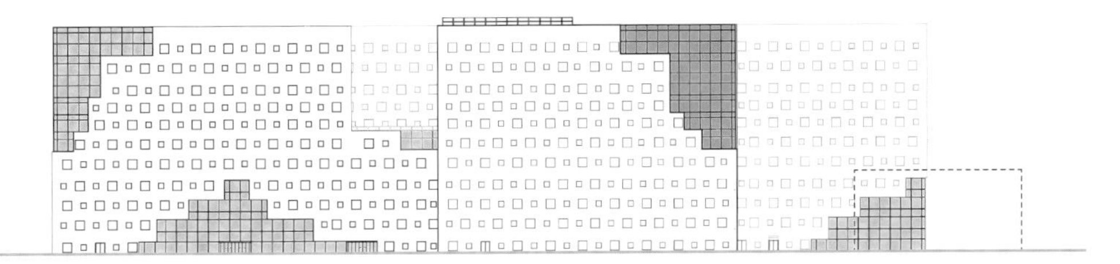

北立面

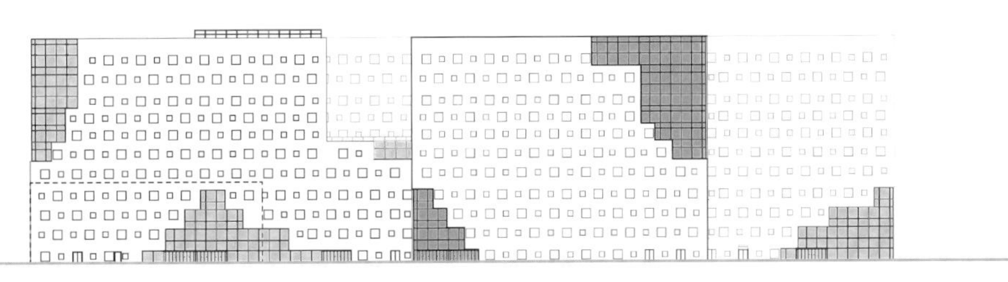

西立面

苏州太湖新城地下空间（中区）

SUZHOU TAIHU NEW TOWN UNDERGROUND SPACE (MIDDLE AREA)

项目类型：轨道地下空间　　**项目地点**：江苏苏州
用地面积：135300m²　　**建筑面积**：148900m²
设计时间：2012年　　**竣工时间**：2019年
合作单位：株式会社日建设计

2022年度江苏省第二十届优秀工程设计地下建筑与人防工程一等奖
2022年度江苏省城乡建设系统优秀勘察设计地下建筑与人防工程一等奖

项目位于苏州太湖之畔的吴中太湖新城核心区中轴大道下方，南临太湖苏州湾，中区与苏州轨道交通4号线支线苏州湾北站无缝连接。该项目包括地下步行街、地下商业、地下停车、人防工程、地面景观和地上观景平台，是TOD站城一体化建设的地下综合体，也是国内首个获得绿色建筑三星级设计标识认证的独立式地下空间项目。

The project is located under the Central Avenue in the core area of Wuzhong Taihu New City on the shore of Taihu in Suzhou. The project is adjacent to Suzhou Bay of Taihu in the south and the middle area is seamlessly connected to Suzhou Bay North Station, a branch line of rail transit line 4, to the north. The project includes underground pedestrian street, underground business, underground parking, human defense project, ground landscape and above ground viewing platform, which is an underground complex of TOD station-city integrated construction, and is also the first independent underground space project in China to be certified with three-star design mark of green building.

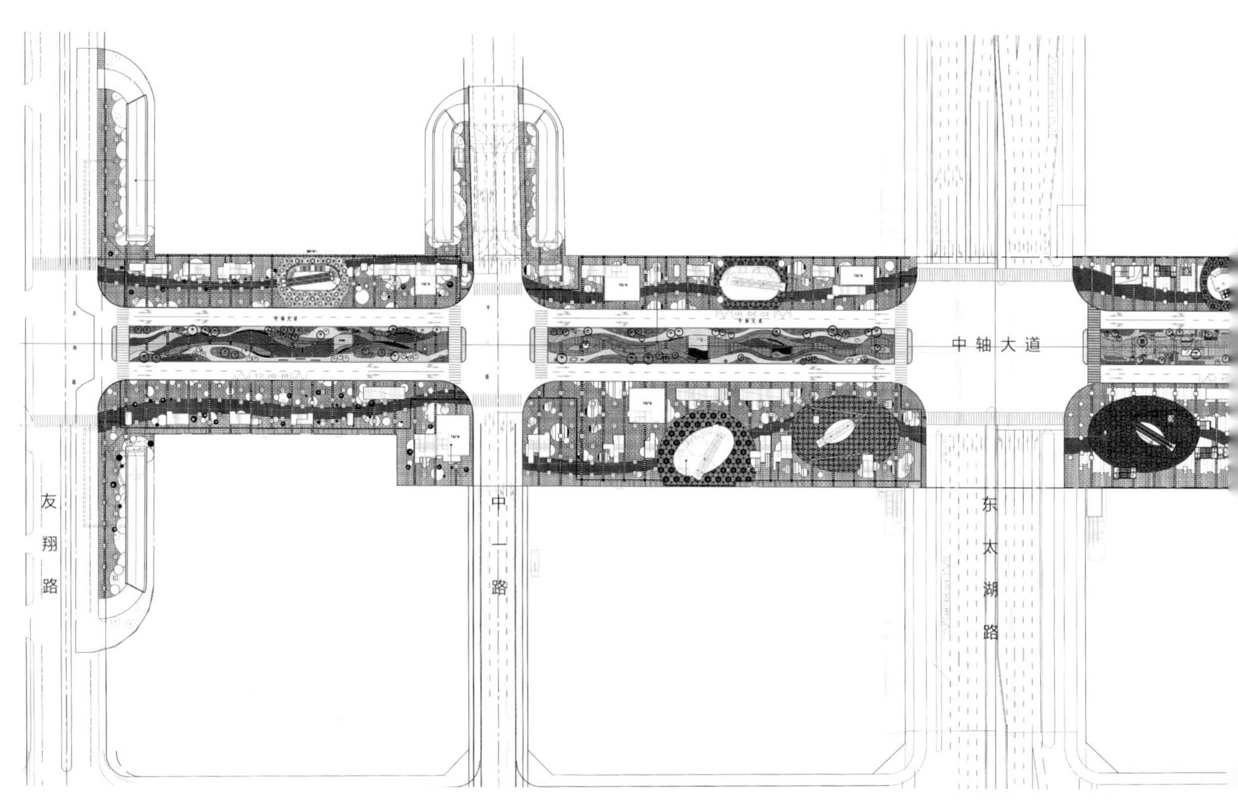

总平面

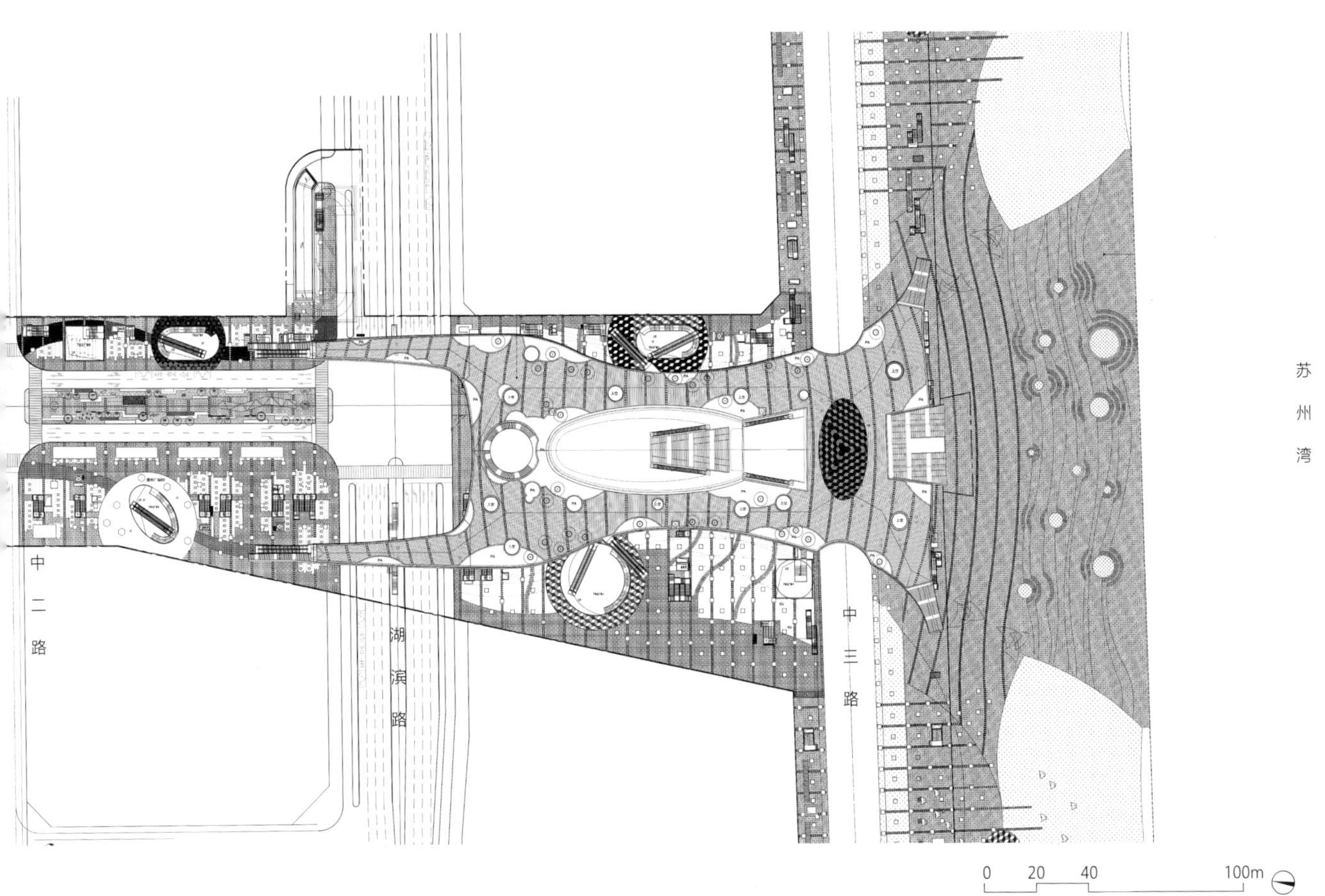
苏州湾
中二路
湖滨路
中三路
0 20 40 100m

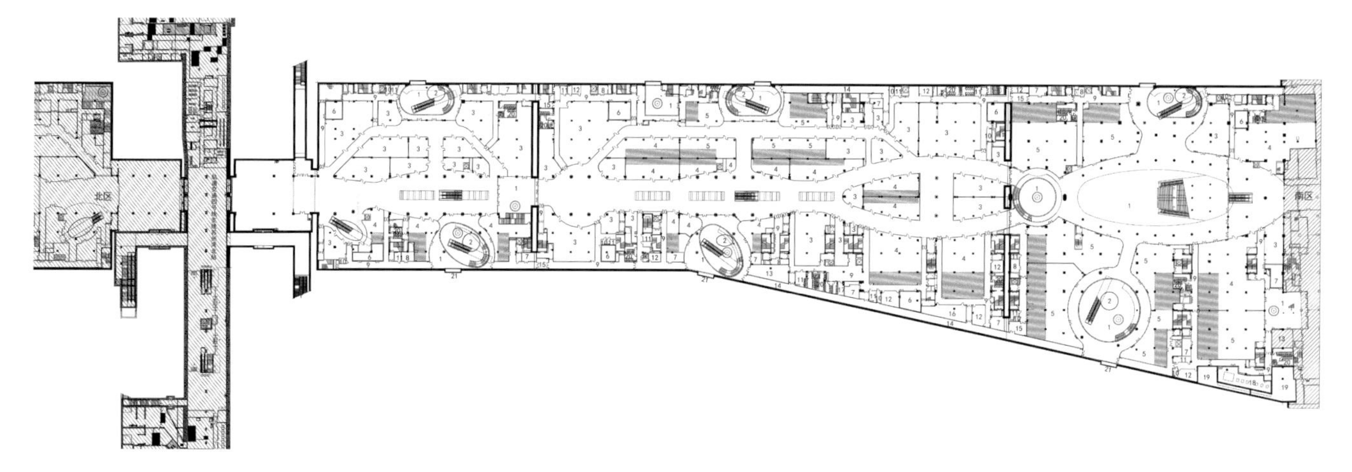

地下一层平面

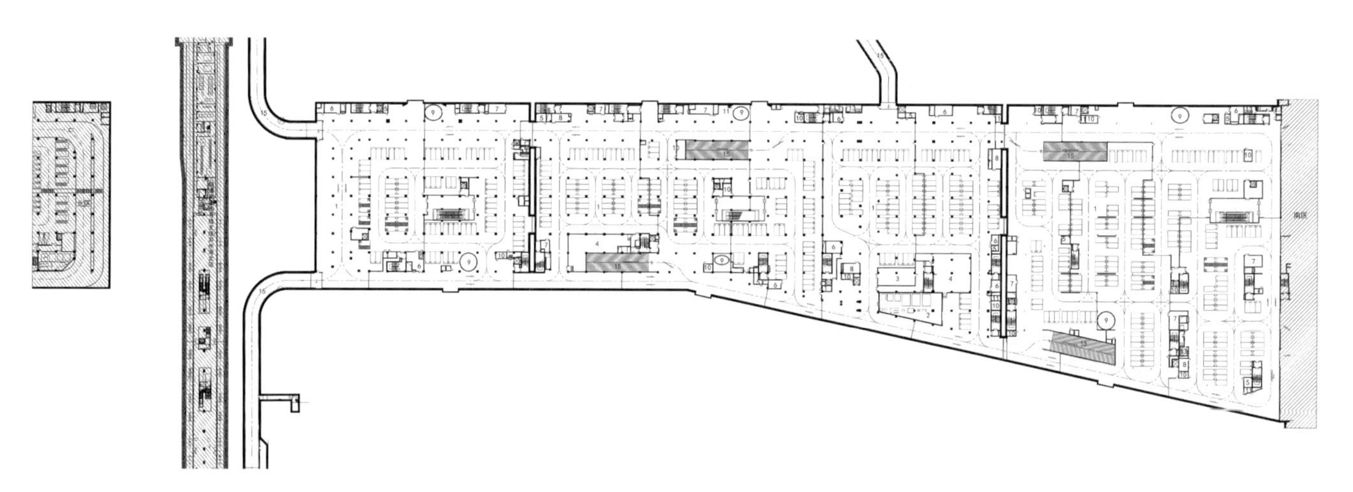

地下二层平面

地下三层平面

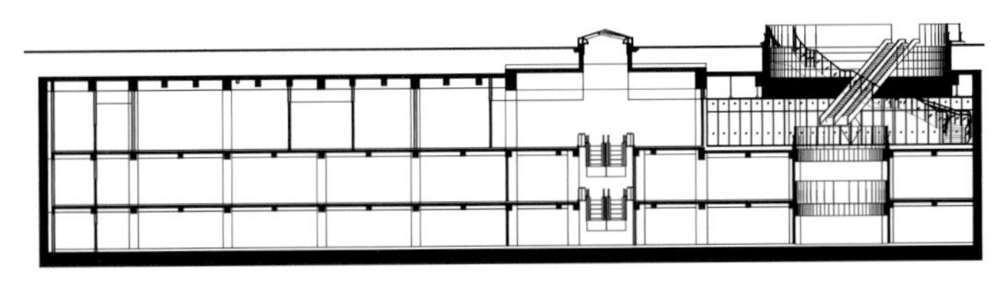

1-1 剖面

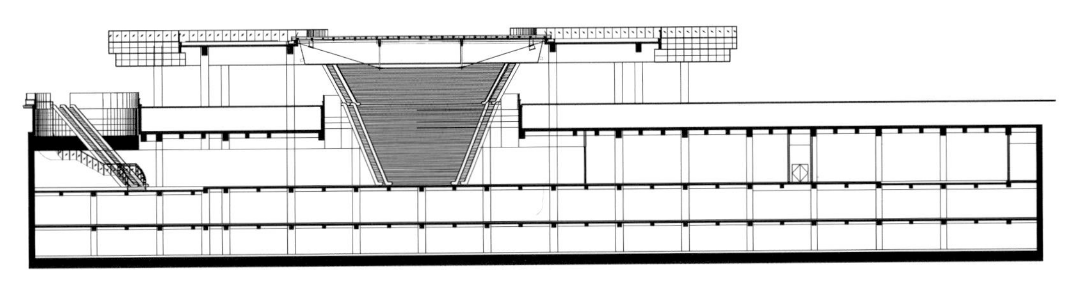

2-2 剖面

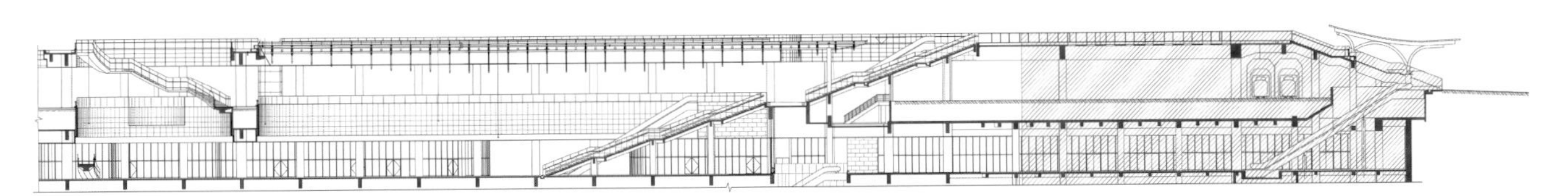

平台剖面

项目以引入太湖自然景观、创建特色地下空间、形成地铁站和湖岸的繁华轴线、连接中心轴和城市的网络体系为设计目标。设计以“天之河”为概念，在地面上设置了星罗棋布的下沉式广场和采光天窗，将自然景观和通风引入地下，极大提升了地下商业空间的舒适感和品质感。南侧太湖大堤处营造的多层褶曲裙摆造型城市观景平台，将核心区南北中轴线慢行交通与东西向城市机动车交通完全分流，更是成为了市民观赏、打卡太湖美景的休闲胜地。

The project is designed to introduce the natural landscape of Taihu, create a special underground space, form a prosperous axis between the subway station and the lakeshore, and connect the central axis with the network system of the city. The design is based on the concept of "River of Sky", with a series of sunken plazas and light skylights on the ground to bring the natural landscape and ventilation underground, greatly enhancing the sense of comfort and quality of the underground commercial space. The multi-layer pleated skirt modeling urban viewing platform on south side of the Taihu embankment can divide completely the core area of the north-south axis of slow traffic and east-west urban motor vehicle traffic. The platform also provides a recreational destination for the public to view and enjoy the beauty of Taihu.

02

办公建筑

OFFICE BUILDING

中新大厦
ZHONGXIN BUILDING
苏州自贸商务中心
SUZHOU FREE TRADE BUSINESS CENTER
苏州港口发展大厦
SUZHOU PORT DEVELOPMENT BUILDING
文旅万和广场
WANHE PLAZA SUZHOU CULTURAL TOURISM GROUP
金融港商务中心
FINANCIAL PORT BUSINESS CENTER
苏州现代服务广场
SUZHOU MODERN SERVICE PLAZA
苏州腾讯数字产业基地
TENCENT DIGITAL INDUSTRIAL BASE OF SUZHOU
招商银行大厦
CHINA MERCHANTS BANK BUILDING
交通银行苏州分行大厦
BANK OF COMMUNICATIONS OF SUZHOU BRANCH BUILDING
苏州银行大厦
BANK OF SUZHOU BUILDING
中国移动园区综合办公大楼
SIP COMREHENSIVE OFFICE BUILDING OF CHINA MOBILE

中新 · 汇金大厦
SIP PROVIDENT FUND MANAGEMENT CENTER BUILDING
苏州国库支付中心
SUZHOU TREASURY PAYMENT CENTER
浙商银行苏州分行
ZHESHANG BANK SUZHOU BRANCH
邮储银行苏州分行
POSTAL SAVINGS BANK SUZHOU BRANCH
吴江软件园综合大楼
COMPREHENSIVE BUILDING OF WUJIANG SOFTWARE PARK
锦峰国际商务广场
JINFENG INTERNATIONAL BUSINESS PLAZA
建屋广场 C 座
GENWAY SQUARE BUILDING C
苏州系统医学研究所
SUZHOU INSTITUTE OF SYSTEMIC MEDICINE
苏州世界贸易中心
SUZHOU WORLD TRADE CENTER
国发平江大厦
GUOFA PINGJIANG BUILDING
国裕大厦二期
GUOYU BUILDING PHASE Ⅱ

中新大厦

ZHONGXIN BUILDING

项目类型：商业办公/超高层　　**项目地点**：江苏苏州

用地面积：12884m²　　**建筑面积**：136574m²

设计时间：2011年　　**竣工时间**：2015年

合作单位：美国GP建筑设计有限公司

2016年度中国建筑学会建筑工程优秀设计奖电气二等奖

2017年度全国优秀工程勘察设计行业奖电气二等奖

2016年度江苏省第十七届优秀工程设计二等奖

2016年度江苏省城乡建设系统优秀勘察设计二等奖

这座222.8m高的标志性大厦，是独墅湖月亮湾地区的制高点。塔楼的中心和独墅湖的圆心点对齐，与该地区的总体规划辐射状视觉效果相符，对城市天际线起着重大影响。北部和南部的主要外墙采用竖向金属凹槽，并用彩陶玻璃带强化塔楼的竖向形态。东西立面采用曲线形的设计，以呼应独墅湖月亮湾湖滨的弧线形状。当夜幕来临，中新大厦便成为月亮湾的灯塔。

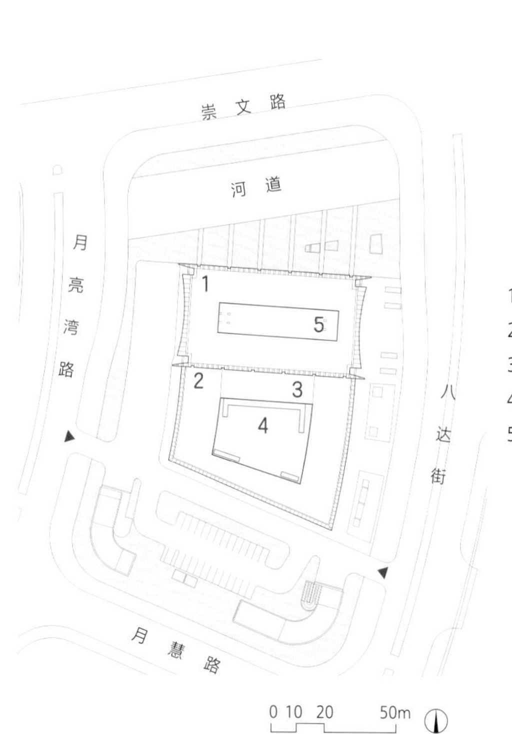

总平面

苏州自贸商务中心

SUZHOU FREE TRADE BUSINESS CENTER

项目类型：城市综合体 /超高层　　**项目地点**：江苏苏州

用地面积：31053m²　　**建筑面积**：263427m²

设计时间：2017年　　**竣工时间**：建设中

合作单位：株式会社日建设计

项目位于苏州园区高铁站正北侧，是继苏州中心之后的又一个超大体量 TOD项目，运用第三代综合体开发理念致力于打造企业总部基地核心区域内高品质办公载体，满足战略性新兴产业对于职能总部、培育平台、孵化平台的高标准办公需求。同时为整个企业总部基地提供酒店、会展、商业等高价值的功能配套项目。

设计以“以无界城境，造开源智桥”为理念，注重功能性、经济性和美观性。将象征丝绸开幕的立面与裙房一体优雅地覆于公园两侧，在丝绸之幕立面之间设计了如同山峦般的立面造型，象征着园区的开拓精神。两栋塔楼整体协调呼应的立面效果更增强了项目作为总部基地核心的标志性和门户性。

建筑外观采用现代化设计风格，结合了玻璃、金属和混凝土等材料，营造出现代感和专业感。裙房造型结合丝绸的意向采用曲线和退台的柔性设计，在表达自身特色的同时，也衬托出高层塔楼的挺拔和大气。

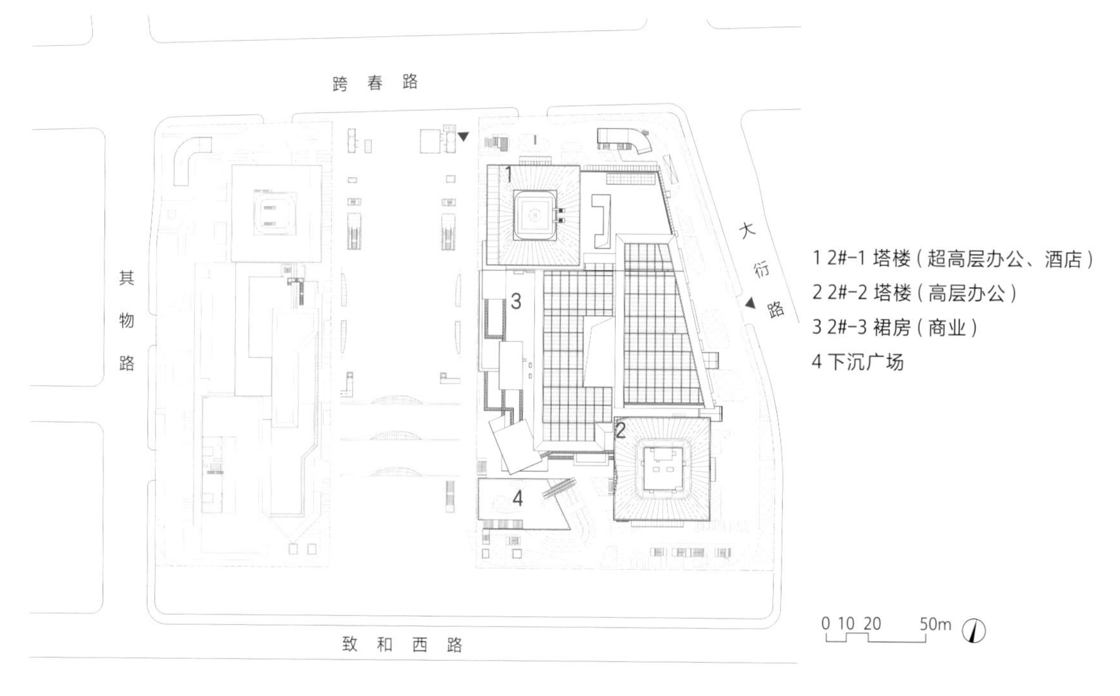

总平面

LOUIS VUITTON
苏州恒泰
HENGTAI HOLDING

苏州港口发展大厦

SUZHOU PORT DEVELOPMENT BUILDING

项目类型： 商业办公/超高层
项目地点： 江苏苏州
用地面积： 10596m^2
建筑面积： 94231m^2
设计时间： 2012年
竣工时间： 2019年

2020年度江苏省第十九届优秀工程设计二等奖
2020年度江苏省城乡建设系统优秀勘察设计奖二等奖

高铁新城作为苏州未来重要的建设区域之一，是为城市提供交通、商业、文化等多功能的服务区。项目位于高铁新城的核心地区，紧邻高铁站南广场，致力于展示高铁新城和苏州城市形象。项目地块大体呈矩形，总体布局按照场地规划要点以及周边道路环境情况，将主要车行出入口置于西侧和南侧，北侧和东侧为主要人行出入口，地下车库出入口位于场地西侧和东南角。

建筑用玻璃体块暗示主要出入口位置，与两边石材形成对比，营造商业氛围，增加了建筑层次感。外立面主要采用银色铝板、浅灰色石材和玻璃三种材质，彰显现代感和科技感。作为一座超高层建筑，竖向线条的使用使得塔楼更显挺拔修长。在塔楼四个立面的正中各设置一道竖向凹槽，为每个立面提供了视觉焦点，同时也赋予了建筑标志性的元素。

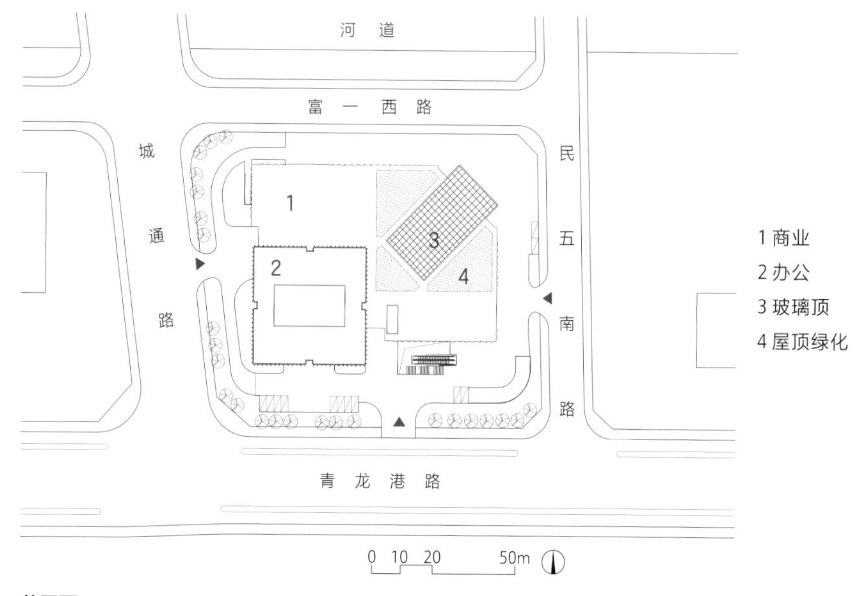

总平面

momenta

文旅万和广场

WANHE PLAZA SUZHOU CULTURAL TOURISM GROUP

项目类型：商业办公/超高层　　**项目地点**：江苏苏州
用地面积：10521m²　　**建筑面积**：81707m²
设计时间：2012年　　**竣工时间**：2016年

2020年度江苏省第十九届优秀工程设计一等奖
2020年度江苏省城乡建设系统优秀勘察设计奖一等奖
2019年度江苏省土木建筑学会第十三届“建筑创作奖”二等奖

项目位于苏州高铁新城商务核心区，定位为集商务会所、休闲娱乐及SOHO公馆于一体的现代城市商务中心。

设计以高铁新城整体规划为依据，力求满足城市设计对该地块的要求。硬朗挺拔的竖向线条与北侧建筑柔和延展的曲线形成对比，相互烘托映衬，强化了各自的特点，再通过两栋建筑塔楼间体块的咬合交错实现调和。通过对比和调和的方式，不仅突出了建筑的个性，也保持了核心区域城市空间形态的整体性。裙楼延续了竖向线条的韵律，通过体块咬合与虚实对比，和主楼构成了有机的整体。建筑西立面体块的悬挑和收分活跃了商业气氛，丰富了西侧城市界面的视觉形象。

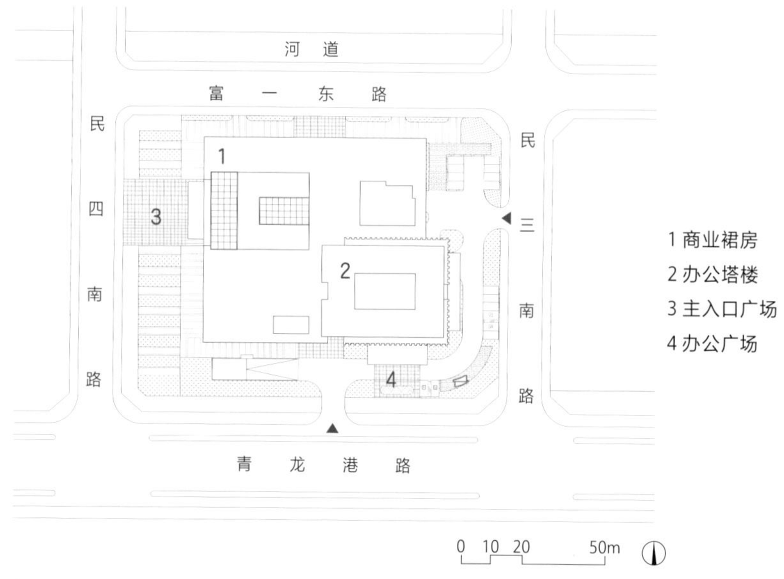

总平面

金融港商务中心

FINANCIAL PORT BUSINESS CENTER

项目类型：商业办公/超高层　　**项目地点**：江苏苏州
用地面积：10521m^2　　**建筑面积**：148500m^2
设计时间：2012年　　**竣工时间**：2016年
合作单位：美国GP建筑设计有限公司

2020年度江苏省第十九届优秀工程设计一等奖
2020年度江苏省城乡建设系统优秀勘察设计奖一等奖
2019年度江苏省土木建筑学会第十三届“建筑创作奖”二等奖

本项目地处苏州重要地段，位于金鸡湖东侧正在建设中的新中央商务区主轴线，由两栋大楼构成。其中，99.8m高的西楼为中信银行苏州总部，一共22层，具备复合功能，包括分行、员工餐厅、营销部、会议/培训设施，以及计算机信息中心等。150m高的东侧超高层主要承担酒店、公寓以及商业办公餐饮等功能。设计采取绿色低碳与智能化理念，优化空间及动线，合理运用各种材料突出功能性。

项目充分考虑了和城市的空间关系以及地域性的独特之处。尤其是每栋塔楼东、西末端的斜切角，打开了望向苏州古城中心和邻近公园的视觉通廊，强化建筑在周边繁华街道上所呈现的动态形象。项目通过灵活而高效的设计，提供优质的办公环境，同时对周边城市环境的影响降全最低，实现了商业价值与社会价值的和谐统一。

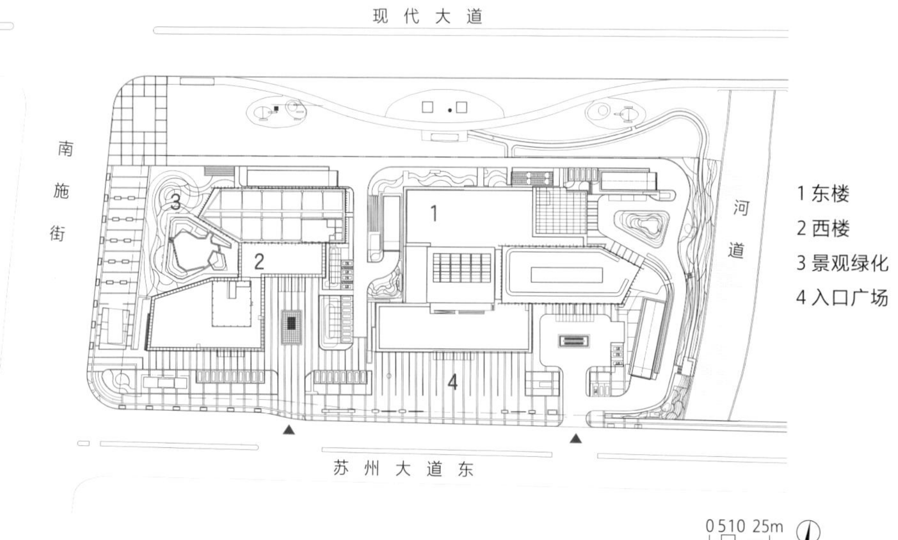

总平面

DOUBLETREE
SBS

SBS
DOUBLETREE

苏州现代服务广场

SUZHOU MODERN SERVICE PLAZA

项目类型：产业园 / 总部办公　　**项目地点**：江苏苏州

用地面积：42600m^2　　**建筑面积**：218000m^2

设计时间：2020年　　**竣工时间**：建设中

2020年度江苏省土木建筑学会第十四届“建筑创作奖”一等奖

项目位于苏州工业园区娄江快速路北侧，距金鸡湖1.5km。项目分为东西两块用地，其中东地块为公寓、商业、办公，西地块为研发办公、研发中心、商业、展览等。项目旨在建设现代服务业的聚集地，推动金融科技、生产性服务业及智能制造产业升级的示范地，聚力构建以“金融科技+中高端生产性服务业+智能制造”为基础的现代服务业大生态圈。

设计将产业生态概念融入园区中。在开放空间和空中平台设计了慢行空间，以促进人际交往，将园区真正变成开放的公园，成为融建筑、自然、人文于一体的第三空间。

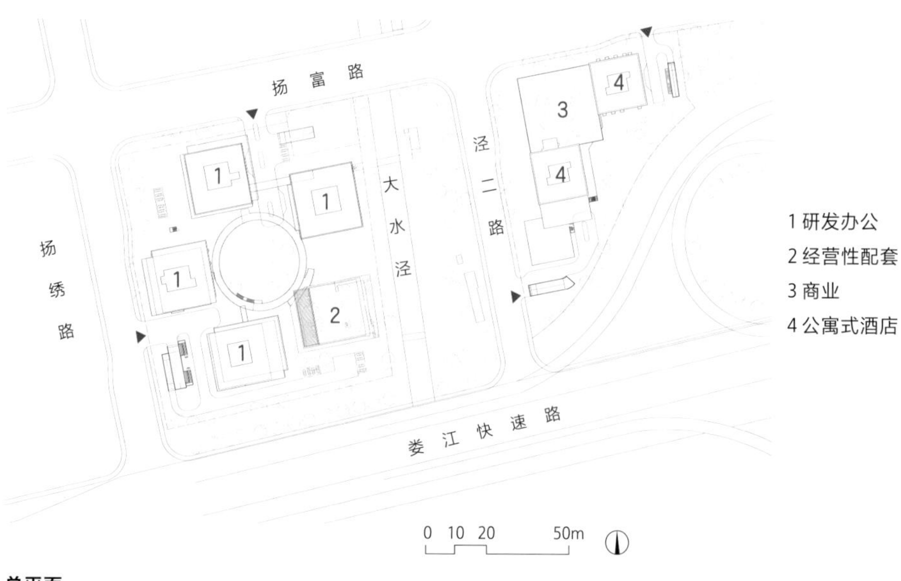

总平面

苏州腾讯数字产业基地

TENCENT DIGITAL INDUSTRIAL BASE OF SUZHOU

项目类型：产业园/总部办公　　**项目地点：**江苏苏州
用地面积：61570m^2　　**建筑面积：**670000m^2
设计时间：2021年　　**竣工时间：**建设中
合作单位：中国建筑西南设计研究院有限公司

项目位于苏州高新区大阳山片区，该片区将集聚大量数字经济企业，全力打造数字产业集聚发展高地。本项目由13栋多高层商务办公楼、一栋高层酒店公寓楼、一栋多层商业以及配套的地下室组合而成。

项目以山水云图超级链接为核心理念，设计建立在自然本底之上，归属于城市进程之中，旨在打造一片绿色生态的城市与自然融合之地、多元创新的数字信息交互之地、智慧科技的产业与生活创新之地、汇聚人气的体验与休闲复合之地。在整个园区中规划了连贯东西的景观带及建筑连廊，结合海绵园区及水生态的科学理念，实现绿色园区设计。建筑造型上采用不规则的多体量组合、穿插、搭砌的设计手法，打造出高品质现代办公园区的形象。

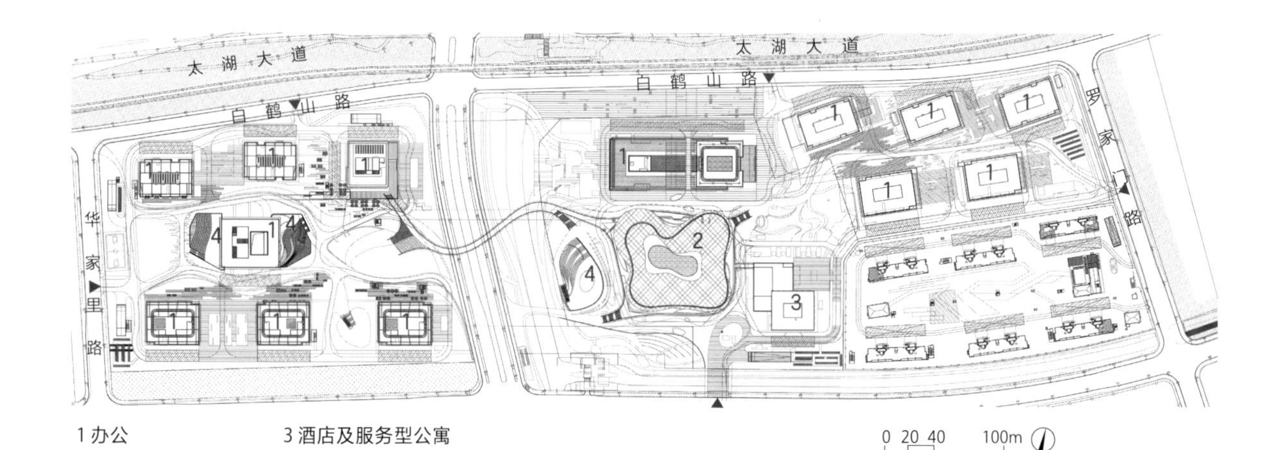

1 办公　　3 酒店及服务型公寓
2 商业　　4 下沉庭院

总平面

Tencent

招商银行大厦

CHINA MERCHANTS BANK BUILDING

项目类型：金融办公
项目地点：江苏苏州
用地面积：7837m²
建筑面积：31972m²
设计时间：2008年
竣工时间：2011年
合作单位：JPW建筑事务所

2013年度全国优秀工程勘察设计行业奖二等奖
2012年度江苏省城乡建设系统优秀勘察设计二等奖
2012年度江苏省第十五届优秀工程设计二等奖

项目位于苏州工业园区湖东CBD区域，为该片区高层建筑群中的一个重要组成部分。建筑在形体上采用裙房加点式塔楼的方式，避免了板式高楼带来的拥堵感。

设计师通过立面体块上坚实与通透的结合在强调永久和实力的同时，也加强了公众参与的重要性，平面的功能空间与立面造型紧密结合，内外渗透。塔楼西侧设计的双层可呼吸式幕墙系统在降低能耗的同时加宽了双层幕墙之间的距离，形成一个可供员工休憩的温室花园，温室花园内连接上下层的楼梯促进了楼层各部门员工间的交流，鼓励有助健康的步行交通方式，同时可减少人们对核心筒电梯的依赖，进一步加强温室花园的共享性。通过对建筑高度与形体的控制，使城市界面统一协调。大厦裙房主要分为营业区和配套辅助功能区。塔楼为行政办公区，采用开放式的大空间办公环境，极具舒适性。整个空间围绕核心筒设置，各单元布置灵活，实用性强。整个建筑外观大方稳重，与周边建筑相互融合，以一种卓然而立的姿态与附近的高层建筑形成呼应。

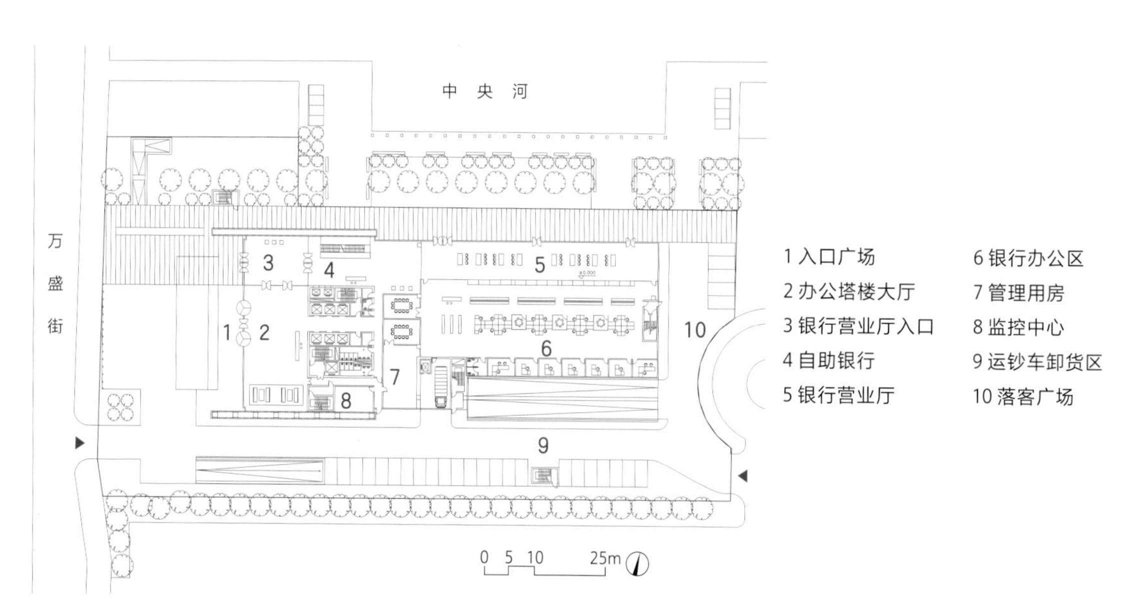

总平面

西北角透视

交通银行苏州分行大厦

BANK OF COMMUNICATIONS OF SUZHOU BRANCH BUILDING

项目类型：金融办公　　**项目地点**：江苏苏州
用地面积：8333m²　　**建筑面积**：64468m²
设计时间：2009年　　**竣工时间**：2013年
合作单位：美国DP设计事务所

2015年度第五届“华彩奖”铜奖
2015年度全国优秀工程勘察设计奖三等奖
2014年度江苏省城乡建设系统优秀勘察设计二等奖

项目位于湖西CBD核心区，由一栋22层的塔楼和与之连接的四层裙房组成。建筑风格借鉴了苏州园林的古典元素和地方文化。它的斜切面转角和以中国传统窗花图案为主题的幕墙是典型苏州园林大门的映射。在塔楼和裙房之间布置景观绿化的开敞内庭院，成为园林空间的经典类型。塔楼的东、西两侧为开放式大办公空间，为灵活使用和自然通风采光提供了条件。东、西立面的锯齿形玻璃外墙与南、北立面规则整齐的窗墙形成对比。塔楼与裙房的屋顶均有屋顶花园，成为员工的休闲场所。

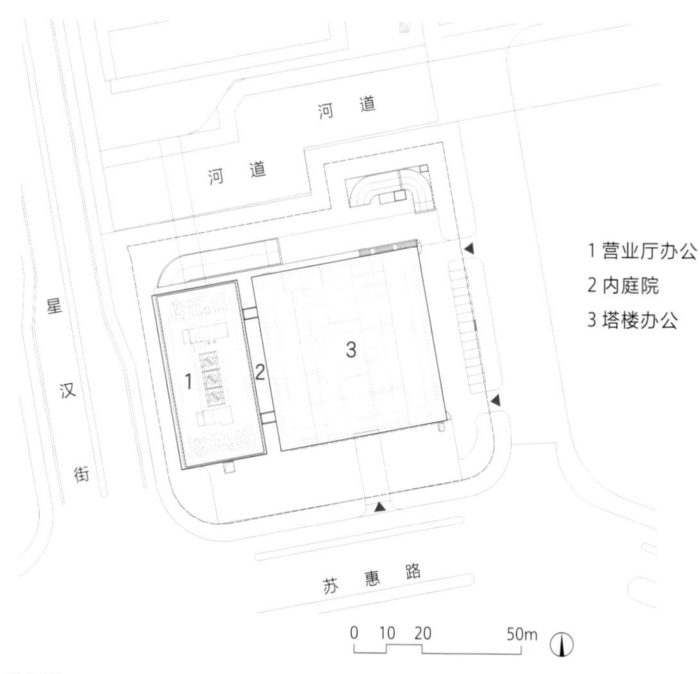

总平面

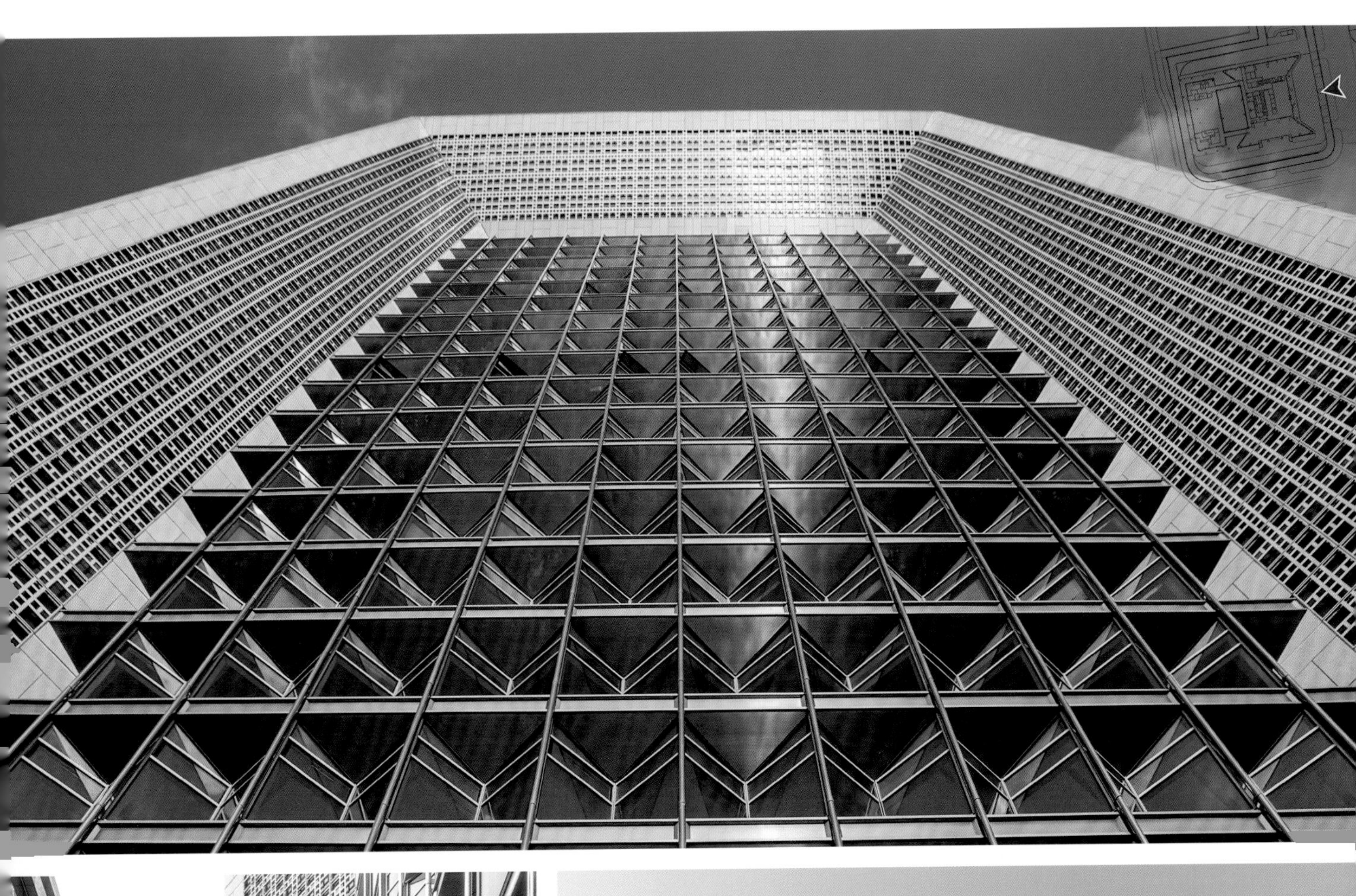

苏州银行大厦

BANK OF SUZHOU BUILDING

项目类型：金融办公　　**项目地点**：江苏苏州

用地面积：13766m²　　**建筑面积**：69588m²

设计时间：2008年　　**竣工时间**：2012年

2015年度江苏省工程勘察设计行业奖电气三等奖

项目位于苏州工业园区综合商贸区核心位置，建筑造型和外部景观、内部庭院、屋顶绿化进行一体化的设计，将苏州园林中的“借景”手法贯穿于整个设计之中。采用了“垂直园林”的设计理念和“庭院盆景”内院式的布局方式，配合主动节能方式，使得建筑更有韵律的同时进一步降低建筑能耗。中心庭院优化了采光和通风效果，同时为使用者营造休憩、交流的场所；裙楼屋顶花园与共享空间相互对景相互渗透，加强了空间的连续性与共享性。

设计通过在建筑体形上用裙房加点式塔楼的方式，放弃板式高楼带来的压迫感，确保城市界面的完整性。采用仪式感和标志性较强的建筑语汇，玻璃幕墙上采用竖向线条，营造挺拔的建筑造型，突出苏州银行的企业形象，并和湖东金融区建筑风格相匹配，幕墙系统光影效果丰富，同时具有很好的遮阳效果。

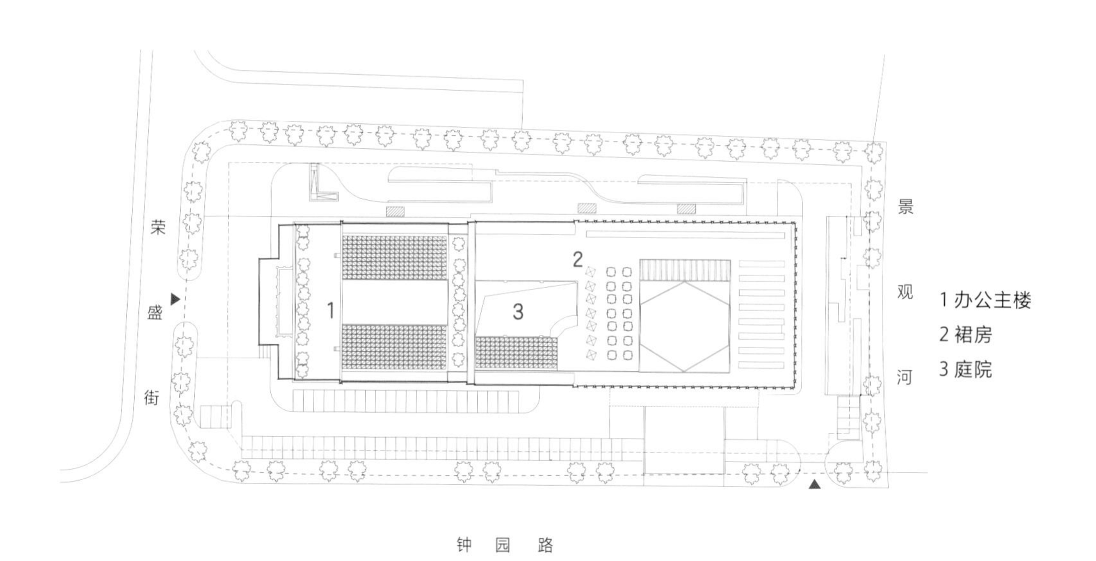

总平面

BSZ
HARMONY

中国移动园区综合办公大楼

SIP COMREHENSIVE OFFICE BUILDING OF CHINA MOBILE

项目类型：金融办公　　**项目地点**：江苏苏州

用地面积：24997m²　　**建筑面积**：85696m²

设计时间：2009年　　**竣工时间**：2014年

合作单位：JPW建筑事务所

2016年度第六届中国建筑学会优秀暖通空调工程设计奖三等奖

2015全国优秀工程勘察设计行业奖建筑工程三等奖

2015年度江苏省城乡建设系统优秀勘察设计奖二等奖

项目位于苏州工业园区金鸡湖东，紧邻苏州工业园区行政中心区。大楼的设计体现了中国移动独特的文化个性，并与周边建筑风格相协调。

设计满足了复杂的功能要求，并将日光引入室内空间。在这个创新决定未来的公司，中国移动希望为员工提供一个高品质的工作环境，因此设计将大楼和绿色景观环境融合在一起。裙房和塔楼设计为一个整体，其中塔楼的设计形成了逐层收进的锥弧形立面，不同面积的楼层适应不同的使用功能。大楼将提供优质的工作环境、灵活的办公空间，以满足业务的需求和未来的发展，成为苏州工业园区未来建筑的新标杆。

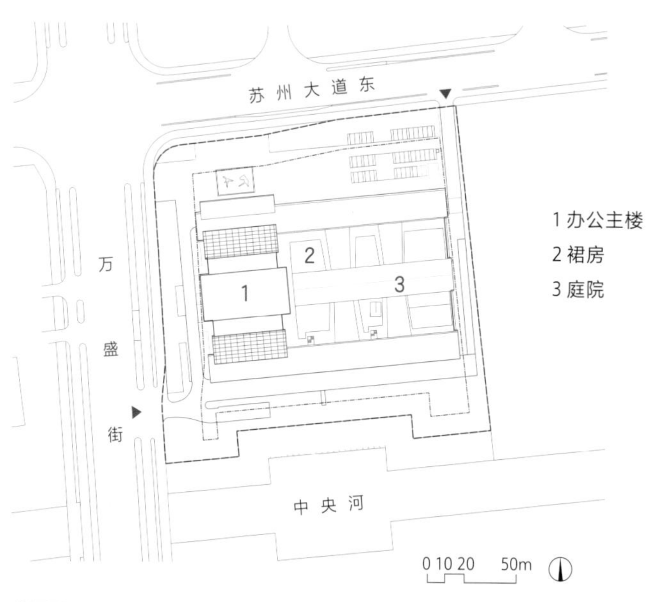

总平面

剖面

中新·汇金大厦

SIP PROVIDENT FUND MANAGEMENT CENTER BUILDING

项目类型：办公　　**项目地点**：江苏苏州
用地面积：14520m²　　**建筑面积**：68440m²
设计时间：2009年　　**竣工时间**：2013年
合作单位：美国GP建筑设计有限公司

2015年度全国优秀工程勘察设计行业奖建筑工程三等奖
2015年度江苏省优秀工程勘察设计行业奖建筑结构三等奖
2014年度江苏省城乡建设系统优秀勘察设计建筑设计一等奖
2014年度江苏省第二十届优秀工程设计二等奖

项目位于苏州工业园区白塘商贸区，是集公积金服务、办公、金融等功能为一体的城市综合体，提供一站式公积金金融服务。项目利用新城市主义的规划思想，以建造对环境友好的节能建筑为设计目标，设计在形体上进行局部的缩进和切削，使建筑本身如同一件精美的雕塑作品，外部幕墙的竖向杆件更是延伸了建筑的立体感，勾勒出一座新颖独特且富有魅力的地标建筑。

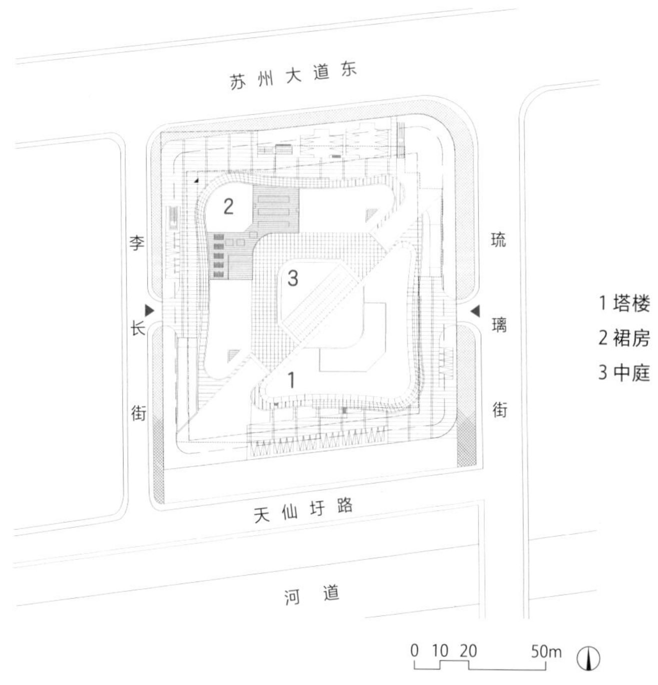

1 塔楼
2 裙房
3 中庭

总平面

苏州国库支付中心

SUZHOU TREASURY PAYMENT CENTER

项目类型：办公　　**项目地点**：江苏苏州

用地面积：19712m²　　**建筑面积**：60741m²

设计时间：2009年　　**竣工时间**：2015年

2017年度全国优秀工程勘察设计行业奖人防三等奖

2016年度江苏省第十七届优秀工程设计地下二等奖

2016年度江苏省城乡建设系统优秀勘察设计地下二等奖

项目位于苏州市政府大院西侧。该项目一期包括两栋办公主楼及一栋会议中心附楼。如何在保留原有建筑群空间关系的基础上，展现新建建筑的特色，同时与周边原有建筑风格统一协调，成为本项目设计的重点。设计借鉴了苏州传统民居及园林的半围合布局方式，两栋高层办公建筑和一栋多层会议综合楼围合成一个院落，与现有办公大楼形成呼应，为办公楼区域营造出丰富的景观环境。

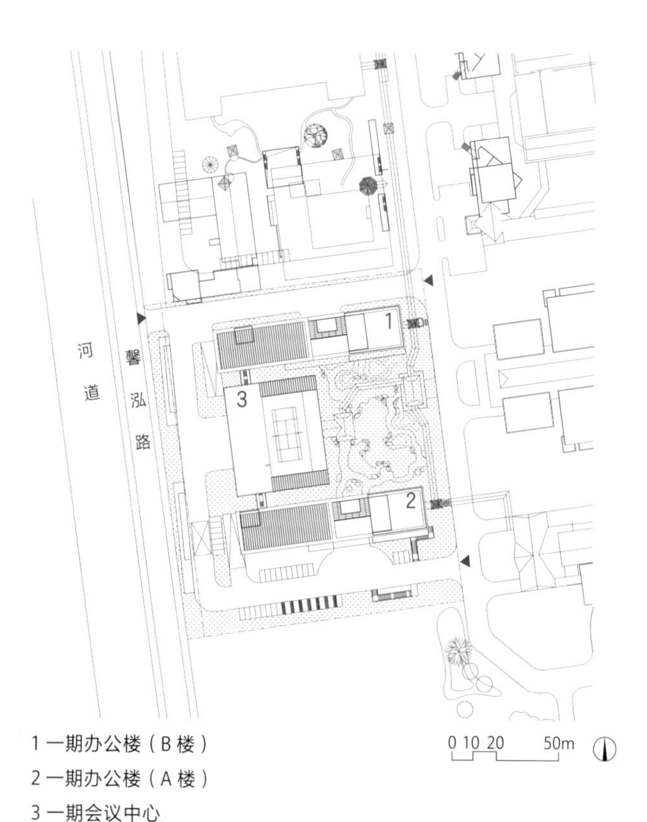

1 一期办公楼（B楼）

2 一期办公楼（A楼）

3 一期会议中心

总平面

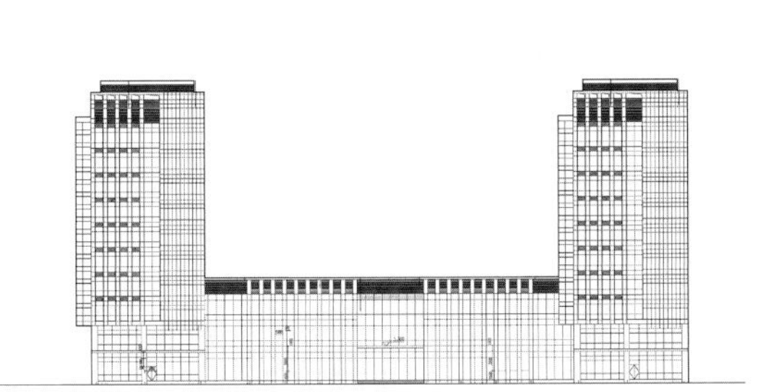

东立面

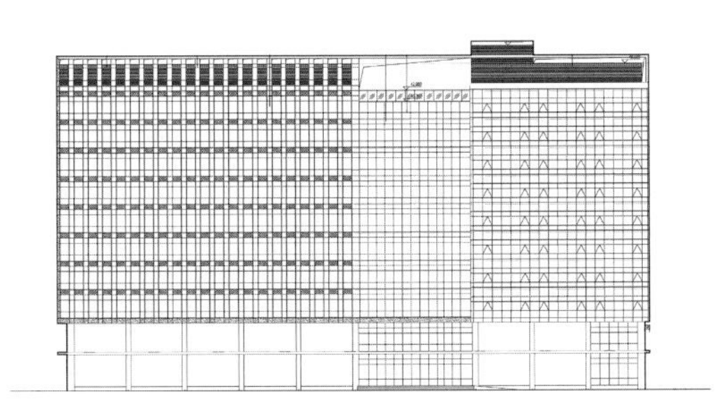

南立面

浙商银行苏州分行

ZHESHANG BANK SUZHOU BRANCH

项目类型：金融办公　　**项目地点**：江苏苏州
用地面积：8687m²　　**建筑面积**：42718m²
设计时间：2017年　　**竣工时间**：2022年

项目位于苏州工业园区旺墩路沿线，总体布局以城市界面和肌理作为切入点，建筑东立面塔楼居于裙房中央布置，形成仪式感和标志性较强的对称界面，以建筑语汇突出浙商银行的企业形象。

裙房通过开敞的中心庭院，为使用者提供休憩、交流的场所，营造公开透明、为客户服务的良好氛围；裙楼屋顶花园与中心庭院互为对景，并与塔楼西侧的休息空间形成互动，加强了空间的连续性与共享性，实现建筑空间富有张力的视野交流。塔楼空间设计引入“垂直园林”的理念，在东侧每隔三层设置一个空中花园，花园设计取自苏州园林意象，内部办公人员身处现代化办公场所，依然能感受小桥流水的悠闲与惬意。

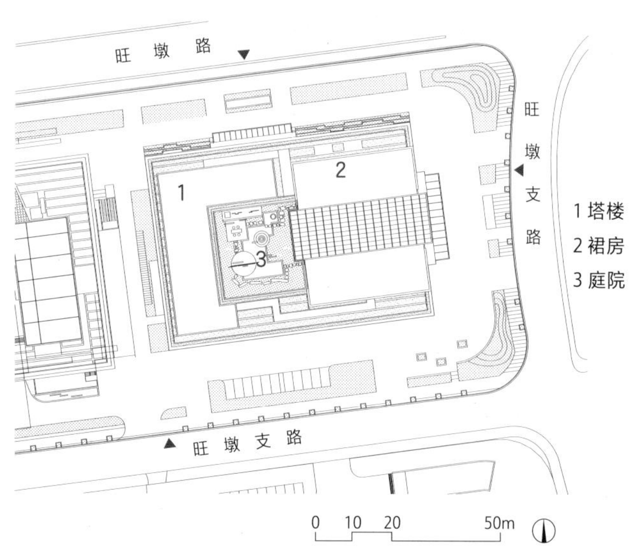

总平面

CZBANK 浙商银行
CZBANK
浙商银行

邮储银行苏州分行

POSTAL SAVINGS BANK SUZHOU BRANCH

项目类型：金融办公　　**项目地点**：江苏苏州

用地面积：10052m^2　　**建筑面积**：47714m^2

设计时间：2017年　　**竣工时间**：2023年

项目位于苏州工业园区旺墩路沿线，周边多为金融办公组群。规划布局顺应主干道东西向布置，既顺应城市界面的延续性，又满足自然采光的生态化。

设计将塔楼置于场地东南角，紧邻主干道钟园路，契合整体城市空间形态，对周边区域的城市界面起到织补作用，同时凸显了自身的建筑特征。裙房与塔楼在形态上相互咬合，用地集约，既满足了不同功能的使用需求，又形成简洁挺拔的建筑体量。建筑形态通过立面重复元素来塑造韵律感，单元式幕墙将线与面相互结合，实与虚形成对比，运用铝板与玻璃的材质，内嵌灯槽勾勒轮廓，来呈现表皮肌理。

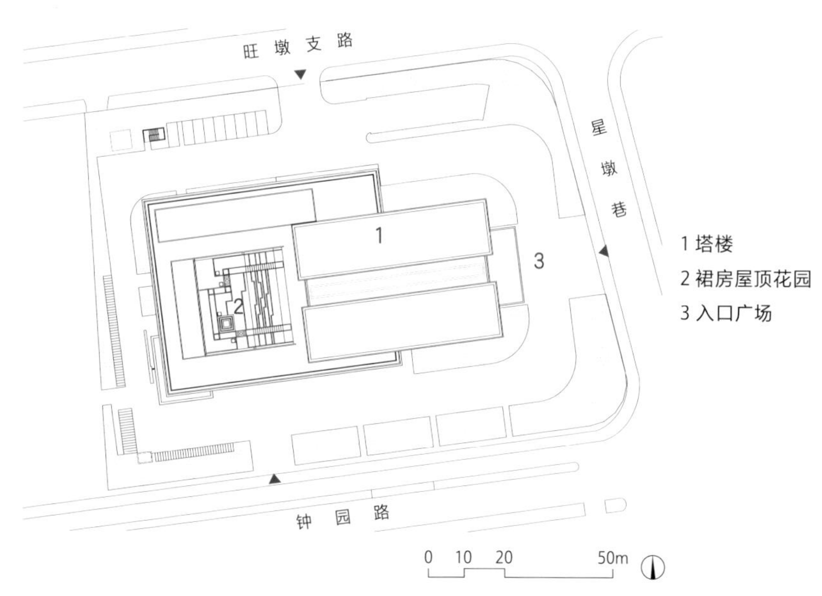

总平面

中国邮政储蓄银行
POSTAL SAVINGS BANK OF CHINA

吴江软件园综合大楼

COMPREHENSIVE BUILDING OF WUJIANG SOFTWARE PARK

项目类型：科研办公　　**项目地点**：江苏苏州
用地面积：9487m²　　**建筑面积**：42014m²
设计时间：2014年　　**竣工时间**：2019年

Architecture MasterPrize Mixed Use Architecture category 2021 HONORABLE MENTION
Architecture MasterPrize High Rise Buildings category 2021 HONORABLE MENTION
2019年度第十三届江苏省土木建筑学会“建筑创作奖”二等奖

项目位于吴江太湖新城，西侧和太湖隔路相望，是一栋容纳湖景、阳光、空气，并将这些资源用于组织软件创业者的工作与生活，帮助他们在各自的空间中思考，在共享空间中碰撞创意的大楼。

项目最大的挑战来自于空间特色和空间收益之间的平衡问题，设计从无序又众多的客观条件中，逐步梳理出对项目有利和不利的条件，渐渐找到了关键问题和打造独特项目属性的可能性。建筑西临宽阔的东太湖岸线美景与繁忙公路，使得湖景与噪声这一对利弊因素的调和，成为了设计难点。与此同时，通透的玻璃幕墙，让西晒问题成为了室内功能布置和节能方面的一个弊端。看湖景还是控噪声？视线通透还是避免西晒？对这些利弊矛盾的妥善处理，成为了设计的切入点和成功的关键。

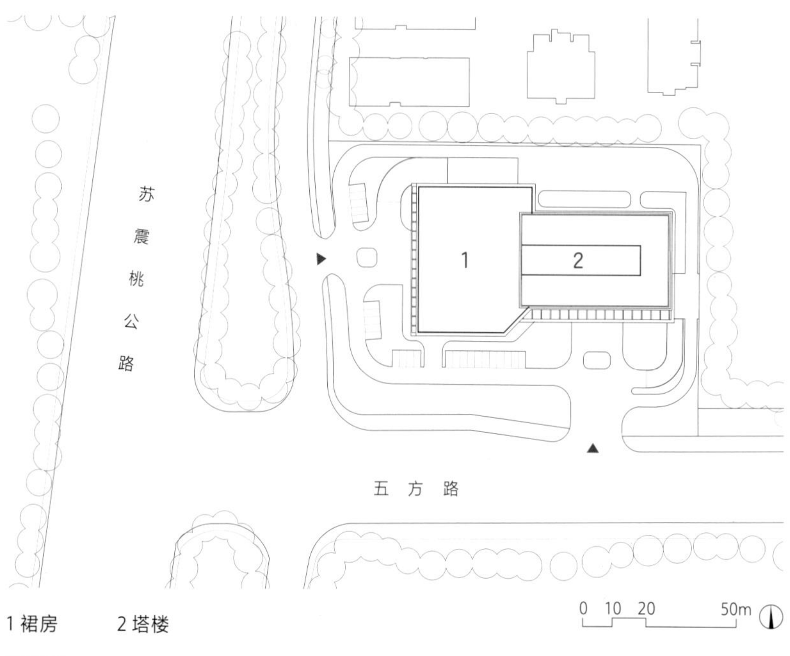

1 裙房　　2 塔楼

总平面

垂直的织补

锦峰国际商务广场

JINFENG INTERNATIONAL BUSINESS PLAZA

项目类型：商业办公　　**项目地点**：江苏苏州

用地面积：30791m²　　**建筑面积**：193058m²

设计时间：2012年　　**竣工时间**：2015年

合作单位：Gensler

2017年度全国优秀工程勘察设计行业奖三等奖

2017年江苏省城乡建设系统优秀勘察设计一等奖

2019年度江苏省第十八届优秀工程设计一等奖

项目位于苏州新区门户，毗邻太湖大道，包括高层甲级写字楼和VIP办公区。设计将主楼与裙房作为一个整体来设计，以一种更有力度、更纯粹的现代主义手法来表现，使之在周围环境中脱颖而出。设计还充分利用周边的自然河景和太湖大道城市绿化带，结合裙房的屋顶绿化，实现自然景观与项目的有机融合。

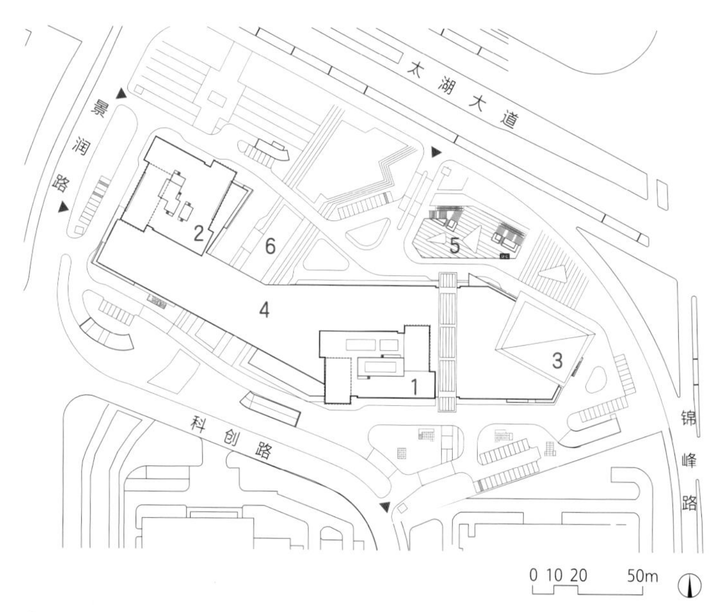

1 A 塔办公楼
2 B 塔办公楼
3 C VIP 办公塔楼
4 办公及配套裙房
5 下沉广场
6 河道

总平面

JFC

建屋广场C座

GENWAY SQUARE BUILDING C

项目类型：总部办公　　**项目地点**：江苏苏州
用地面积：12150m²　　**建筑面积**：76764m²
设计时间：2013年　　**竣工时间**：2017年
合作单位：孚提埃（上海）建筑设计事务所有限公司

2020年度江苏省第十九届优秀工程设计一等奖
2019年度江苏省城乡建设系统优秀勘察设计奖一等奖
2020年度江苏省优秀工程勘察设计行业奖绿建二等奖

项目位于独墅湖高教区西，是月亮湾城市副中心的重要高层建筑，丰富了月亮湾地区的天际轮廓线；东北侧是体育馆和影剧院，北侧临近湖滨公园。设计积极打造开放与共享的交流空间，把丰富的自然景观和城市生活引入建筑中。高层塔楼采用三角形建筑外观形态，在与周边既有建筑协调的同时，又独具特色。设计借鉴苏州特色的传统街巷，营造古典韵味空间。全水景的办公空间结合庭院和屋面绿化，形成完整的绿色体系，使建筑更人性化。

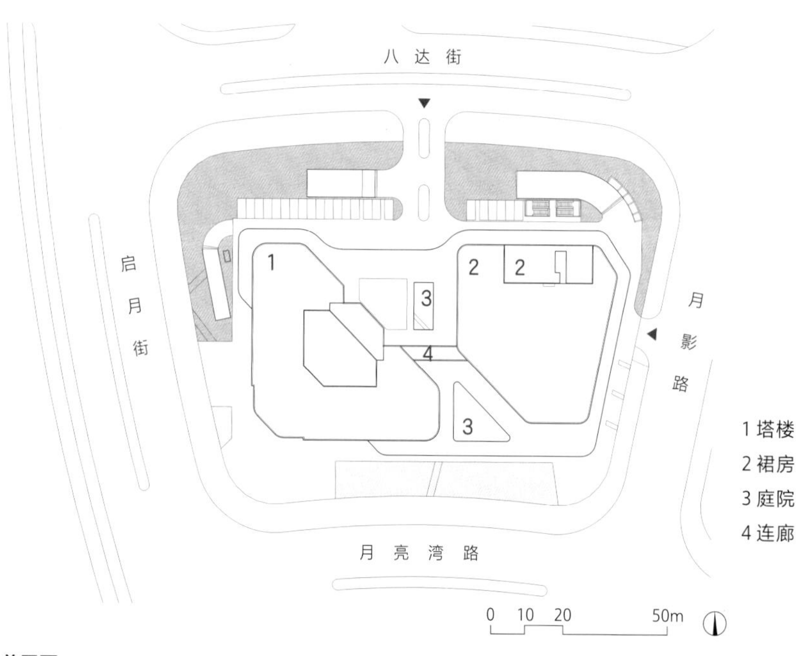

总平面

新建元
SUNGENT

苏州系统医学研究所

SUZHOU INSTITUTE OF SYSTEMIC MEDICINE

项目类型：总部办公　　**项目地点**：江苏苏州

用地面积：12150m^2　　**建筑面积**：76764m^2

设计时间：2013年　　**竣工时间**：2017年

2019年度江苏省城乡建设系统优秀勘察设计奖一等奖

2020年度江苏省第十九届优秀工程设计二等奖

项目位于苏州工业园区科教创新区，南侧为苏州国际科技园，其他三面紧邻苏州大学独墅湖南校区。设计理念来源于细胞链和生命繁衍的灵感，总体布局模仿植物生长的形态，由主干生出枝叶，枝叶的排布模仿细胞链的连接方式自由分布。建筑立面采用简洁大气的单元幕墙设计，大小两种宽度模数的开窗形式互相搭配，严谨又不失活泼。黑白搭配的外墙色彩穿插绿色金属竖向线条，从城市不同角度呈现不同的立面效果，展现了现代高科技园区的风采。

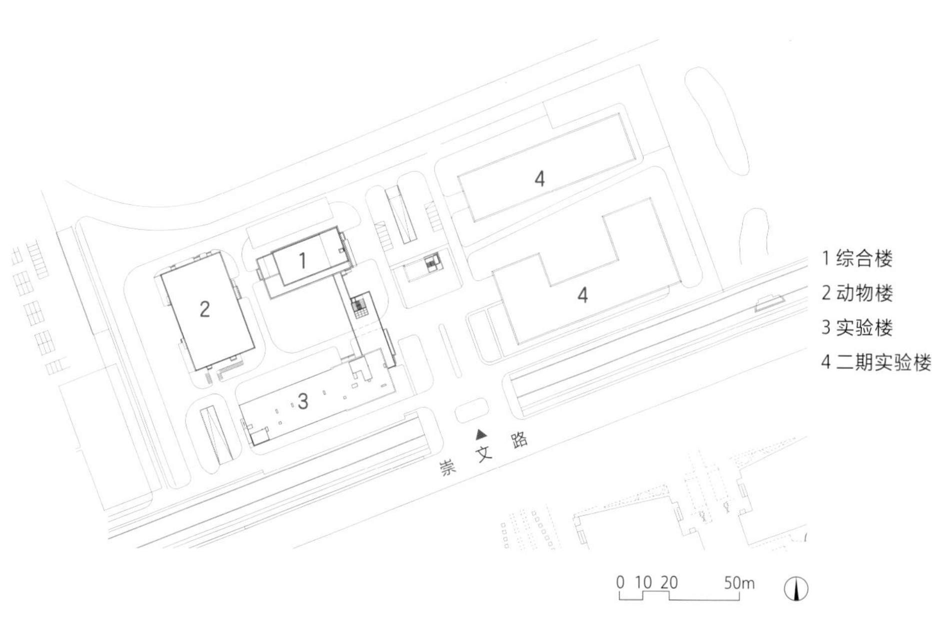

总平面

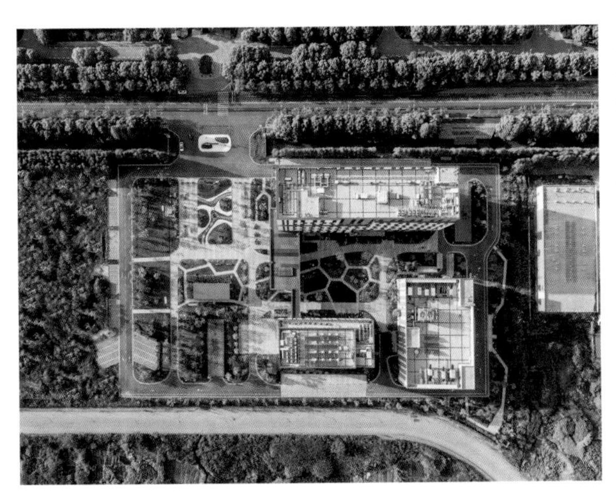

ISM

苏州世界贸易中心

SUZHOU WORLD TRADE CENTER

项目类型：商业办公　　**项目地点**：江苏苏州

用地面积：14999m^2　　**建筑面积**：115544m^2

设计时间：2012年　　**竣工时间**：2016年

2017年度全国优秀工程勘察设计行业奖电气三等奖

2019年度江苏省第十八届优秀工程设计三等奖

2017年度江苏省城乡建设系统优秀勘察设计建筑设计三等奖

2017年度江苏省优秀工程勘察设计行业奖电气一等奖

项目位于苏州火车站以北，隔苏站路与苏州新火车站遥相对望，是城市对外交流和展示城市形象的门户之一，也是区域内的标志性建筑和主要的人流集散空间。项目以“吴门”为设计主题，将建筑分为东西两部分，分别为办公和酒店功能，两者之间通过空中连廊相连，形成整体大气而又富有城市特征的舞台背景。酒店和办公楼的主要出入口均布置在基地北侧的苏站北路上，与建筑南侧的商业人流有效分离。玻璃幕墙的外侧设置了菱形遮阳构件，在形象上与火车站形成紧密的呼应关系。

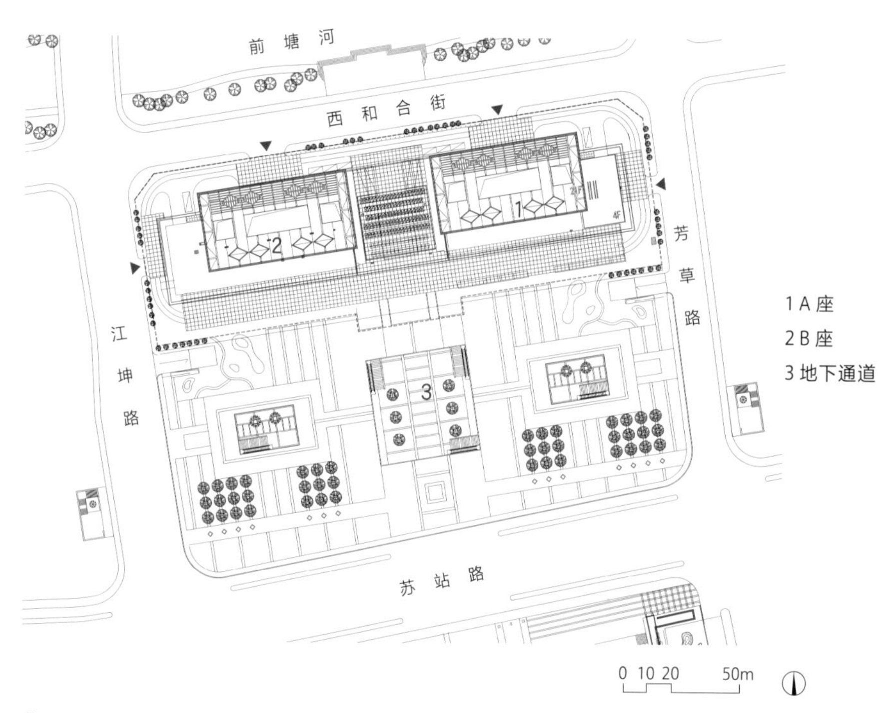

总平面

国发平江大厦

GUOFA PINGJIANG BUILDING

项目类型：商业办公　　**项目地点**：江苏苏州
用地面积：19631m²　　**建筑面积**：115533m²
设计时间：2010年　　**竣工时间**：2014年

2016年度江苏省第十七届优秀工程设计三等奖
2016年度江苏省城乡建设系统优秀勘察设计建筑设计三等奖

项目位于苏州市平江新城，建设用地处在新城与古城控高变化的过渡位置。用地北面地块限高100m，南面地块限高24m和60m，城市形态界面的落差较大，因此两栋塔楼被设计成一高一矮，屋顶轮廓形成阶梯式的斜线，屋顶标高由100m降到98m，再由84m降到76m再到66m，制造一个软过渡，让城市主界面的天际线更加整齐顺畅。

项目在绿色节能方面也进行了深入设计，塔楼西立面设计了独特的三层幕墙系统，南立面设计了光伏发电遮阳板，集遮阳和可再生能源利用的功能于一身，体现了绿色建筑的设计理念。

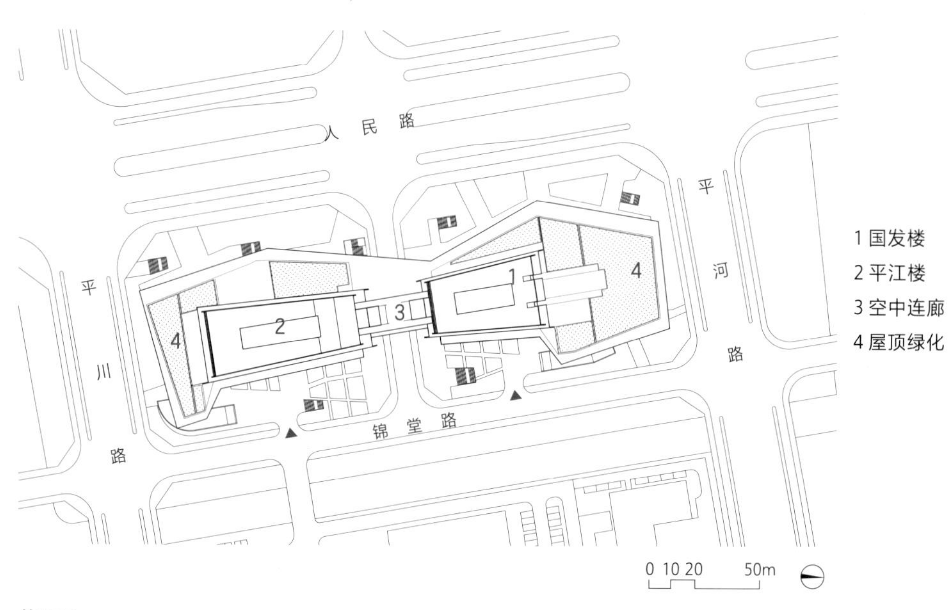

总平面

万融国际

国裕大厦二期

GUOYU BUILDING PHASE Ⅱ

项目类型： 商业办公　　**项目地点：** 江苏苏州
用地面积： 8274m²　　**建筑面积：** 52290m²
设计时间： 2015年　　**竣工时间：** 2018年

2020年度江苏省第十九届优秀工程设计二等奖
2020年度江苏省城乡建设系统优秀勘察设计建筑设计二等奖

本案的目标在于充分结合周边环境，合理利用各项技术手段，提升商务办公建筑形象，营造庄重、简洁的办公环境，实现实用与美观、高效与舒适的统一。设计旨在提供一种自由的工作模式、一个宜人的工作场所。通过设置屋顶花园等措施，将绿色与景观引入建筑内部，与办公空间紧密结合，提升环境的健康与舒适度，使员工身心愉悦，从而提高工作效率。

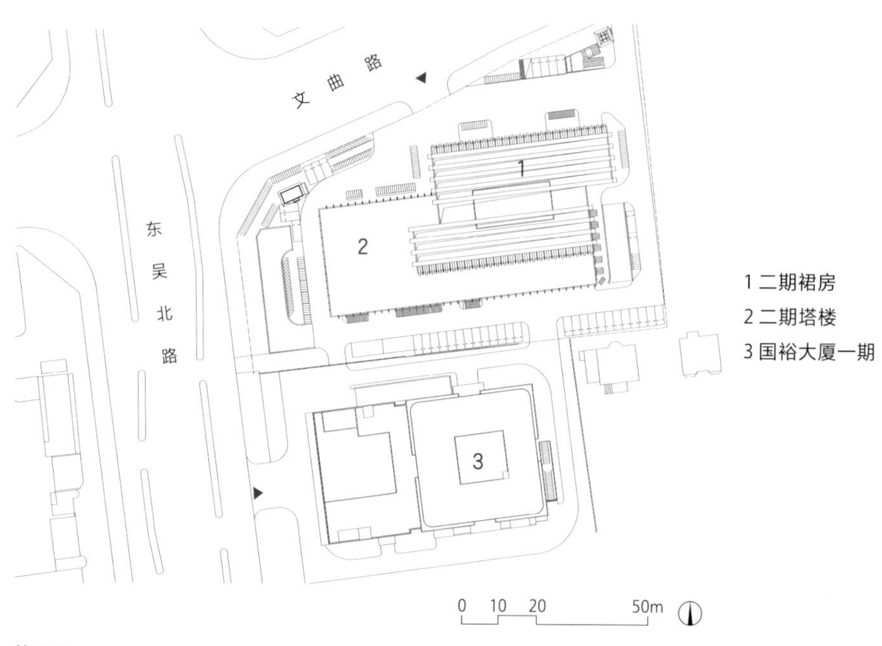

总平面

03

商业综合体

COMMERCIAL COMPLEX

苏州中心
SUZHOU CENTER
丰隆城市中心
FENGLONG CITY CENTER
南通圆融广场
NANTONG HARMONY CITY
元和活力岛城市副中心
YUANHE HUOLI ISLAND URBAN SUB CENTER
苏悦广场
SUYUE PLAZA
恒宇广场二期
HENGYU SQUARE PHASE Ⅱ
月光码头
MOONLIGHT WHARF
龙湖狮山天街
LONGFOR SHISHAN TIANJIE
常熟永旺梦乐城
CHANGSHU AEON
沧浪新城社区服务中心
CANGLANG NEW CITY COMMUNITY SERVICE CENTER

苏州中心

SUZHOU CENTER

项目类型：商业办公　　**项目地点**：江苏苏州

用地面积：167000m²　　**建筑面积**：1130000m²

设计时间：2011年　　**竣工时间**：2017年

合作单位：株式会社日建设计

贝诺建筑设计咨询（上海）有限公司

2018年度中国建筑学会建筑创作奖电气一等奖

2019年度全国优秀工程勘察设计行业奖电气一等奖

2019年度全国优秀工程勘察设计行业奖人防二等奖

2020年度中国建筑学会建筑设计奖城市设计三等奖

2020年度中国建筑学会建筑设计奖结构三等奖

2019年度江苏省第十八届优秀工程设计一等奖

2018年度江苏省城乡建设系统优秀勘察设计一等奖

2018年度江苏省优秀工程勘察设计行业奖电气一等奖

2019年度江苏省优秀工程勘察设计行业奖人防一等奖

项目位于苏州工业园区金鸡湖西侧，居苏州市域CBD核心位置，包括高层塔楼七栋、大型商业建筑一座，集商业、办公楼、公寓、酒店等多种业态，是提升城市品质、完善城市功能的重大基础设施和民生工程，为国内首批“站城一体”的超大型城市共生体。

项目采用“统一策划定位、统一规划建设、统一运营管理”的模式，对地面建筑和地下空间进行整体开发，统筹安排业态布局，统一协调建筑风格，建筑与环境相融、文化与经济共生，全力打造有别于传统综合体的新形态，形成辐射带动城市经济、孕育城市活力的“城市共生体”，建成后将协同周边项目形成苏州市域CBD核心，全面助力园区产业发展与转型升级。

建筑设计配合绿化退台与景观大平台紧密衔接，塑造独特的城市绿洲及绿色环保CBD形象。建筑内部功能分区清晰，利用建筑设计协调高效商业布局，打造特色商业空间，整体规划符合科学、经济、美观的要求。

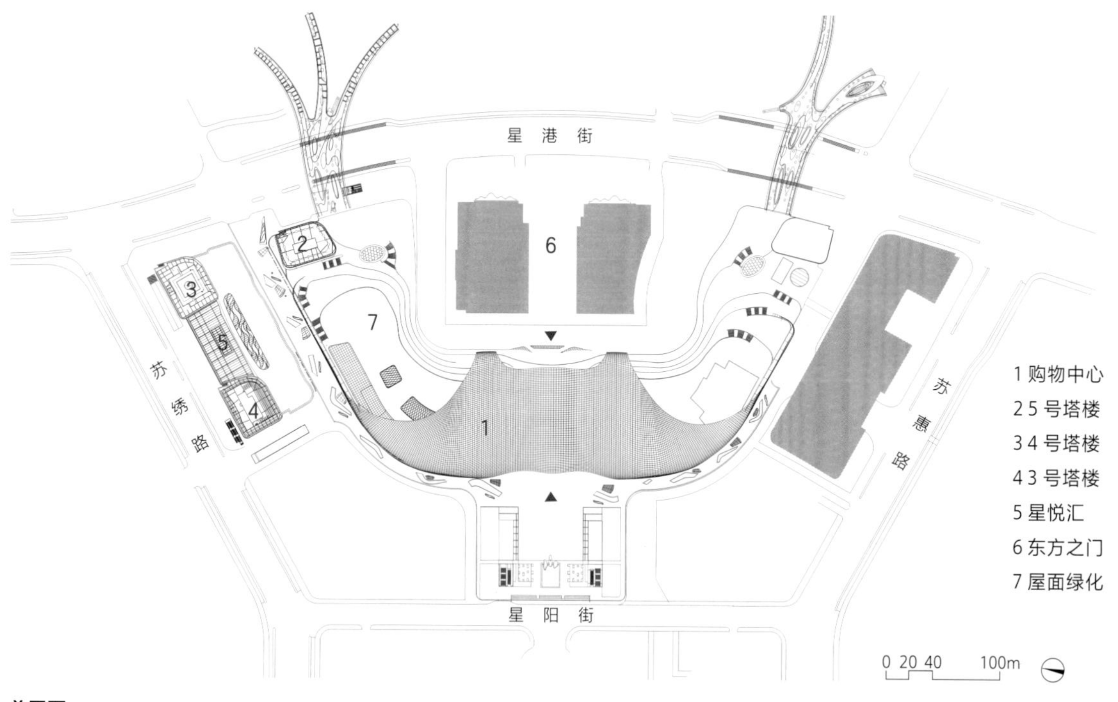

总平面

丰隆城市中心

FENGLONG CITY CENTER

项目类型：城市综合体　　**项目地点**：江苏苏州
用地面积：45455m^2　　**建筑面积**：410987m^2
设计时间：2012年　　**竣工时间**：2018年
合作单位：凯达环球建筑设计咨询（北京）有限公司

2020年中国建筑学会建筑设计奖电气二等奖
2019年度全国优秀工程勘察设计行业奖水系统三等奖
2020年度江苏省第十九届优秀工程设计一等奖
2019年度江苏省城乡建设系统优秀勘察设计奖一等奖
2019年度江苏省优秀工程勘察设计行业奖水系统一等奖
2020年度江苏省优秀工程勘察设计行业奖电气一等奖
2019年度江苏省优秀工程勘察设计行业奖人防二等奖

项目位于金鸡湖畔，主体为四栋塔楼，基地被30m宽的南北向号观大道分隔，四栋塔楼分置四角。为获得最大的景观朝向，塔楼朝向扭转了一定角度，从而能够最大化利用金鸡湖的开阔视野，也为基地预留了发展空间。其中两栋塔楼通过底部大型商业裙房相连，打造为整个地块的商业中心。商业、办公区域均有独立的出入口，互不干扰又方便联系。结合底层局部架空和外观造型的设计，使高大的建筑与城市有机地结合在一起，保持了城市整体空间环境的协调，通过屋面绿化和对设备的屋面格栅进行设计，达到净化城市第五立面的效果。

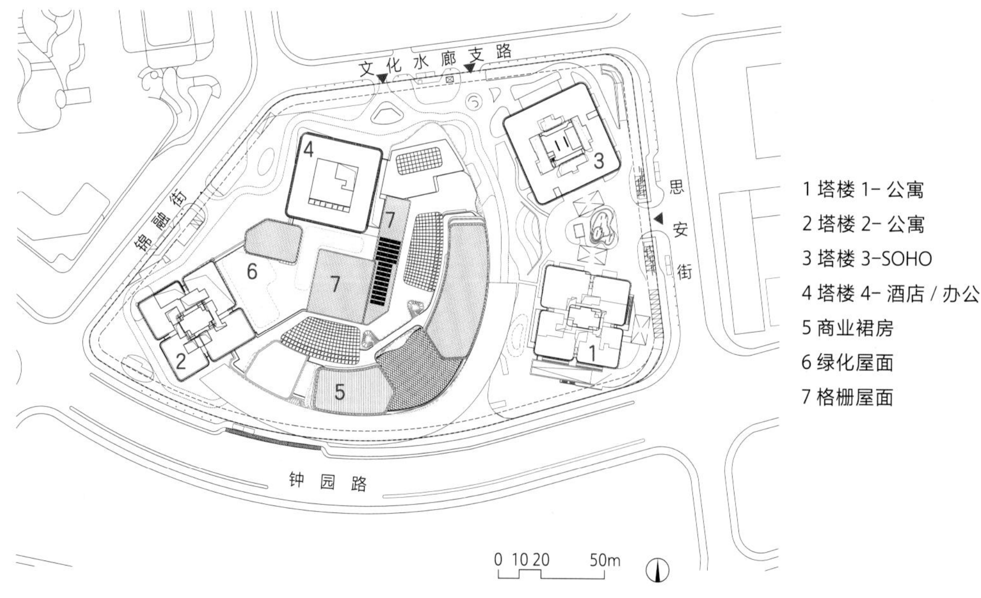

总平面

IFS

南通圆融广场

NANTONG HARMONY CITY

项目类型：城市综合体　　**项目地点**：江苏南通
用地面积：47142m²　　**建筑面积**：307498m²
设计时间：2012年　　**竣工时间**：2017年
合作单位：上海悉地工程设计顾问股份有限公司

2019年度江苏省优秀工程勘察设计行业奖电气一等奖
2019年度江苏省优秀工程勘察设计行业奖结构三等奖

项目位于南通城市南北中轴线交汇的黄金地段，是新老城区的交汇点。项目功能为公寓式酒店、办公和商业。设计构思来源于“钻石永恒之美”，结合钻石切割的设计美学体现圆融广场的地标价值与高端定位。通过对三角“钻石”单元的组合变幻，形成建筑的设计亮点。广场大屋面由若干三角“钻石”单元拼合而成，渐变的色彩犹如钻石各个角度折射的光影；流动的形体又柔化了“钻石”自身的坚硬感。它们或透明或镂空，飘逸灵动、轻舞飞扬。夜幕降临，一片片“钻石”宛若繁星点点，高雅而时尚。“钻石”屋面延伸至建筑各个入口，既可作为遮风避雨的设施，又能提供大面积的城市公共空间。

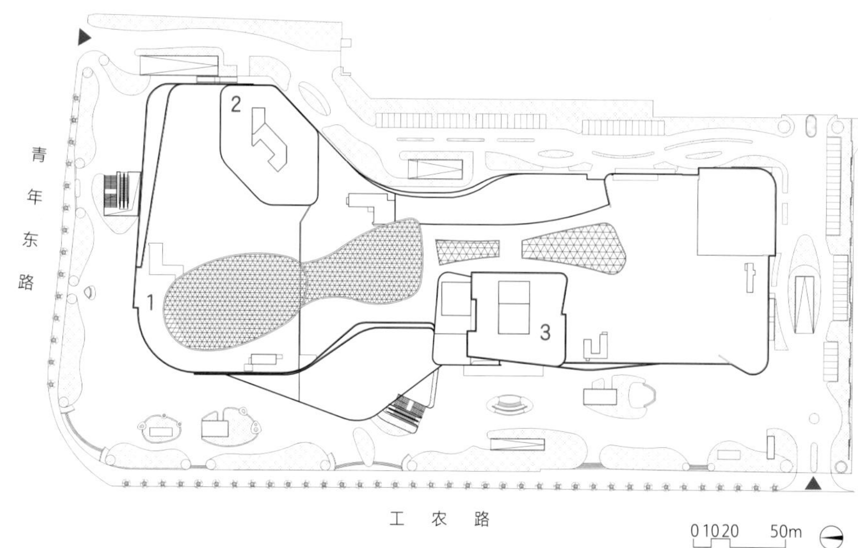

总平面

金鹰

元和活力岛城市副中心

YUANHE HUOLI ISLAND URBAN SUB CENTER

项目类型：商业综合体/城市更新　**项目地点**：江苏苏州
用地面积：83053m^2　**建筑面积**：42496m^2
设计时间：2019年　**竣工时间**：2021年

2022年度江苏省第二十届优秀工程设计二等奖
2022年度江苏省城乡建设系统优秀勘察设计二等奖
2021年度江苏省土木建筑学会第十五届“建筑创作奖”一等奖

项目位于苏州相城中心商贸城的几何中心，是姑苏古城人民路北延的重要空间节点，亦是相城元和高新区东西向生态廊道的中心节点。

提升改造时正逢全国城市发展转向城市更新新语境的关键期，在此背景下，项目以城市更新视角为切入点，从人的根本需求出发，围绕如何重新激发副中心的产业、文化、生态、运动活力，通过城市脉络疏导、多元功能复合、建筑轻型介入、景观生态叠加等设计策略，重新激活城市文脉，提升活力岛凝聚力，使得副中心重焕新生。

设计立足城市中心体系中重要生活型服务中心的规划定位，以重塑片区活力为目标，构筑“活力水城新引擎”“苏州中轴新地标”“生态绿岛新标杆”，提出了“通脉”“理脉”“梳脉”的三大应对策略。

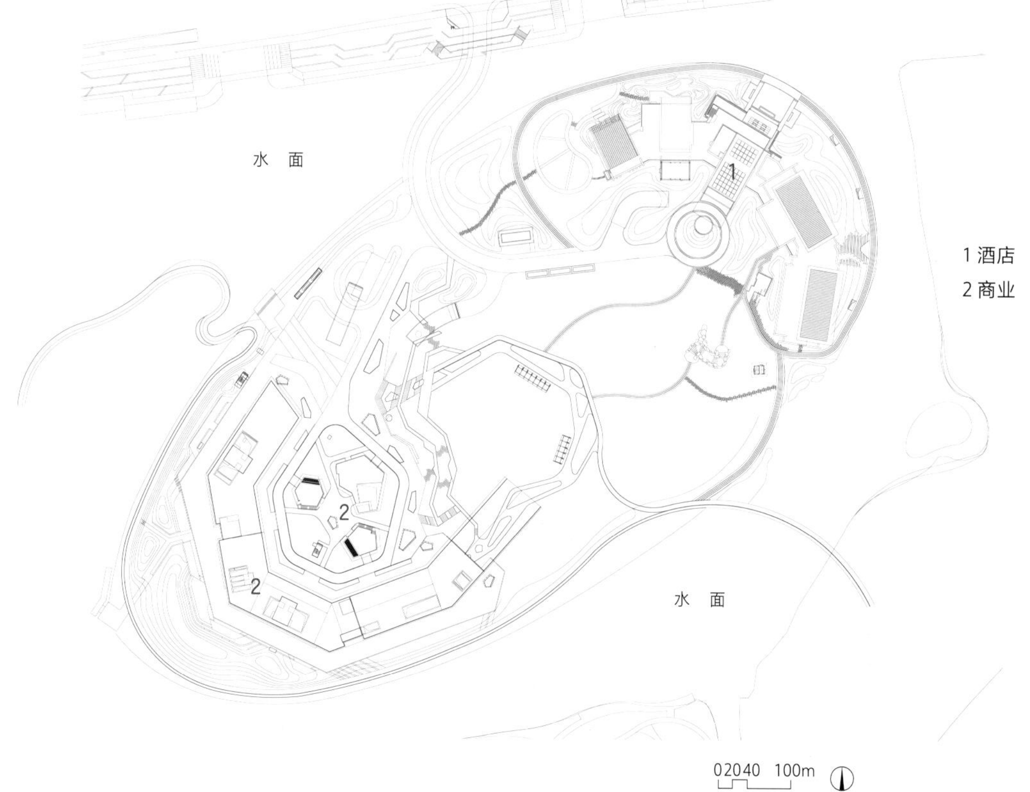

总平面

苏悦广场

SUYUE PLAZA

项目类型：城市综合体　　**项目地点**：江苏苏州
用地面积：15392m²　　**建筑面积**：221575m²
设计时间：2012年　　**竣工时间**：2016年
合作单位：美国GP建筑设计有限公司

2017年度全国优秀工程勘察设计行业奖电气三等奖
2017年度江苏省城乡建设系统优秀勘察设计一等奖
2019年度江苏省第十八届优秀工程设计二等奖
2017年度第四届（BIM）应用设计大赛BIM拓展应用奖
2017年度江苏省优秀工程勘察设计行业奖电气二等奖
2019年度江苏省优秀工程勘察设计行业奖智能化二等奖

项目位于苏州工业园区CBD核心位置，南北两栋塔楼分别高157.15m和82.60m。两栋塔楼使用三层挑高的60m跨无柱连廊，轻盈通透，联系南北楼的商业空间。建筑整体风格简洁大气，与周边建筑保持了较好的协调性。项目采用了多项绿色建筑技术，获得了LEED金奖。

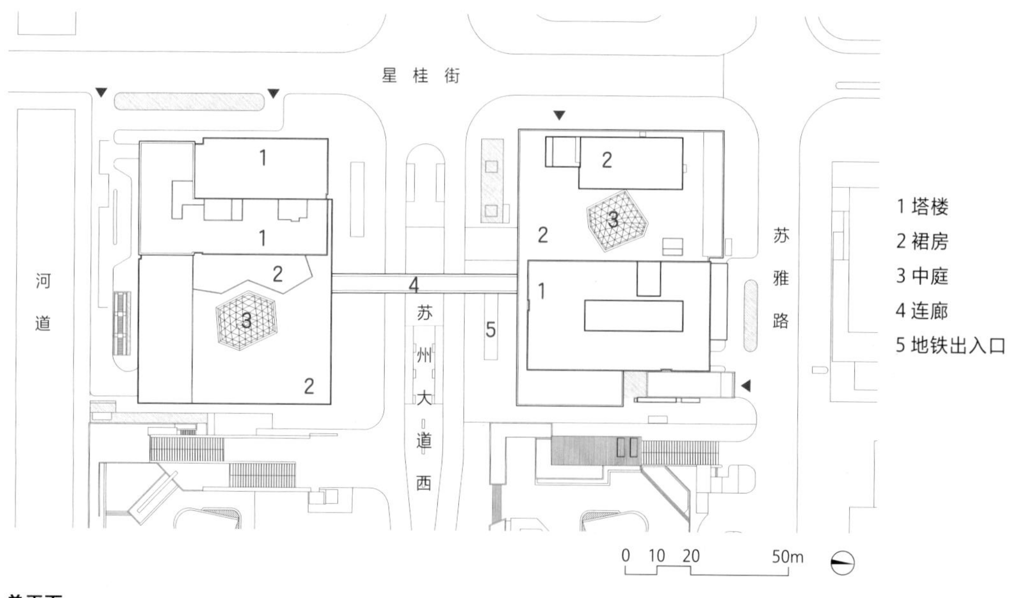

总平面

AIA

恒宇广场二期

HENGYU SQUARE PHASE Ⅱ

项目类型：商业办公　　**项目地点**：江苏苏州
用地面积：8539m²　　**建筑面积**：56148m²
设计时间：2010年　　**竣工时间**：2013年
合作单位：美国GP建筑设计有限公司

2015年度全国优秀工程勘察设计奖三等奖
2015年度江苏省城乡建设系统优秀勘察设计奖二等奖

项目位于工业园区核心区，北临苏绣路，南临苏雅路，是一栋多功能高层综合体建筑。塔楼1-15层为办公，16层为会所，17-26层为精装修公寓。各功能区域出入口相对独立。裙房与塔楼形成了一个围合感很强的内院。为避免幕墙过于平整单调，采用了每层错位每两个单元板块一组逐渐突起的做法，使得外立面层叠有致，建筑随光线的移动不断变化，精致而有趣地映射着城市的繁华景象。

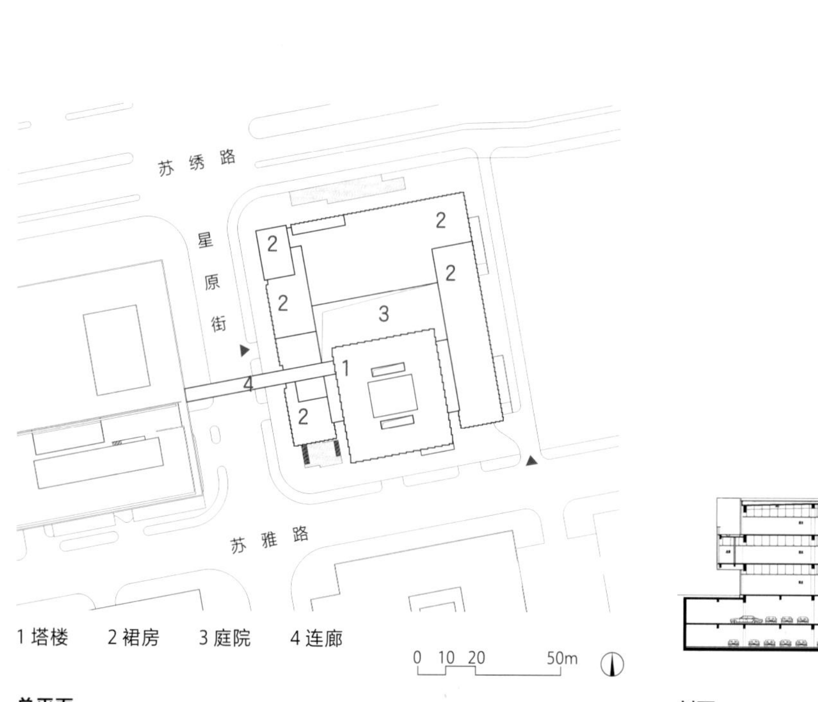

总平面

剖面

月光码头

MOONLIGHT WHARF

项目类型：商业街区　　**项目地点**：江苏苏州
用地面积：45500m²　　**建筑面积**：27768m²
设计时间：2007年　　**竣工时间**：2008年
合作单位：美国优联加建筑设计事务所

2010年度江苏省第十四届优秀工程设计三等奖
2010年度江苏省城乡建设系统优秀勘察设计三等奖

月光码头欧式商业街位于苏州工业园区金鸡湖东北岸，西面与科技文化艺术中心广场相接，北面接苏州国际会展中心广场，东面邻诚品居所。该项目由高级会所、量贩KTV、异国风情餐饮、时尚酒吧等业态的十多栋中小型建筑组合而成，其中在东、西两端各布置一栋主力店。

如何充分利用基地的景观和滨水优势，以“夜天堂”为核心理念，形成以休闲、娱乐、主题餐饮为特色的苏州园区环金鸡湖新商圈，是设计的关键。设计以纯正欧式风格为主，充分发掘项目的临水优势，采用“穿水堤岛”的设计理念，将金鸡湖水引入到基地内部，有机地将基地分割成岛屿或半岛的地形特征，大幅度增加了基地的沿湖岸线，创造了更多的临水和观景空间，打造出一个具备浓郁欧陆浪漫气息的高端休闲娱乐场所。项目整体采用“一条景观主轴”、“两条特色街巷”和“三个空间段落”的布局，营造出多层次的欧式浪漫气质的滨水商业环境。

基地北侧苏州国际会展中心呈现出“扇子”的形象，西侧科技文化艺术中心体现了“珍珠”的意象，本项目的主体商业街区则如绽放的花朵，静静地开放在金鸡湖畔，结合月光码头那一弯新月，体现出“花好月圆”的概念意象。

龙湖狮山天街

LONGFOR SHISHAN TIANJIE

项目类型：商业综合体　　**项目地点**：江苏苏州

用地面积：52734m²　　**建筑面积**：278062m²

设计时间：2013年　　**竣工时间**：2017年

合作单位：上海拜肯建筑设计咨询有限公司

2019年度江苏省第十八届优秀工程设计一等奖

2018年度江苏省城乡建设系统优秀勘察设计一等奖

2019年度江苏省优秀工程勘察设计行业奖结构二等奖

2019年度江苏省优秀工程勘察设计行业奖人防二等奖

基于对总体定位布局、交通流线及商业导向的研究，结合退台设计，为商业打造良好的硬件环境。建筑设计以“一轴一环”来打造整体商业的主动线，内部中庭连接平面空间与垂直空间，每层的中庭都进行差异化打造，提供场景化的游园式商业体验。

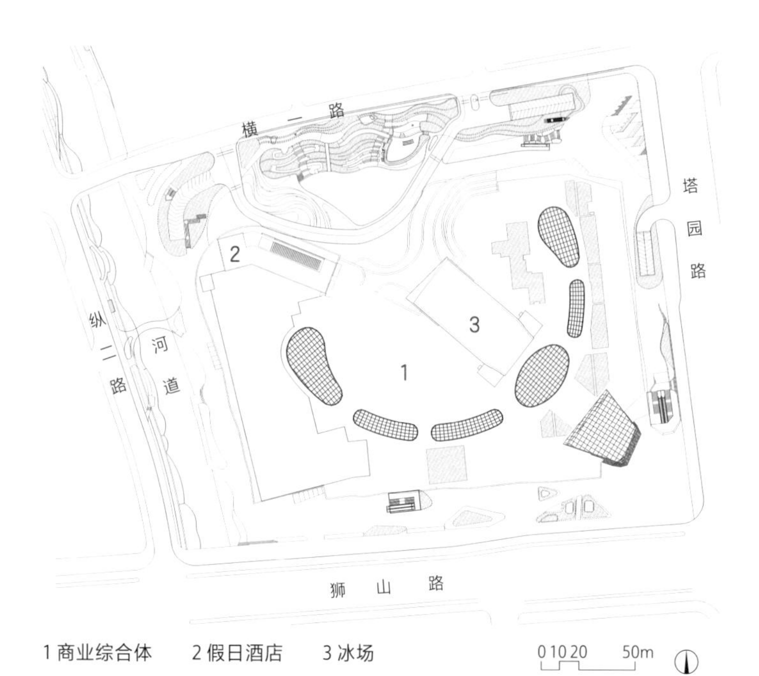

1 商业综合体　2 假日酒店　3 冰场

0 10 20　50m

总平面

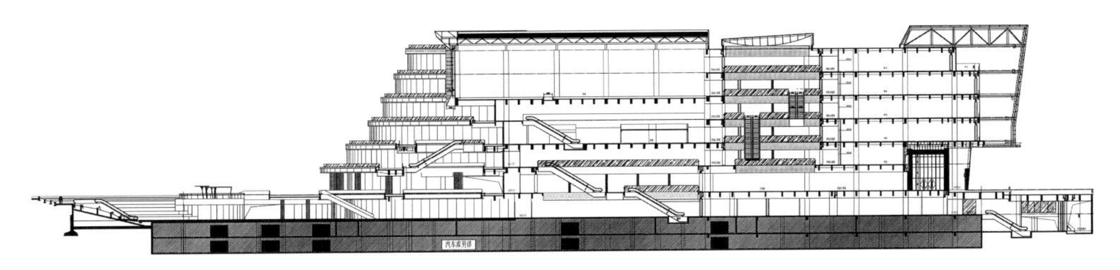

1-1 剖面

常熟永旺梦乐城

CHANGSHU AEON

项目类型：商业综合体　　**项目地点**：江苏常熟
用地面积：125843m²　　**建筑面积**：180898m²
设计时间：2016年　　**竣工时间**：2018年
合作单位：上海船场建筑装饰有限公司

2020年度江苏省第十九届优秀工程设计三等奖
2020年度江苏省城乡建设系统优秀勘察设计奖三等奖
2019年度第六届江苏省勘察设计行业建筑信息模型（BIM）应用大赛三等奖

项目位于常熟市高新区，延续了永旺梦乐城的典型布局，将分布在两端的核心店铺通过一条专卖店街连接起来，形成“两核一街”形式的购物中心，同时按照商铺走向布置中庭。采光带不仅活跃了内部空间，同时有效引导顾客视线，从而引导人流在水平方向穿行，激活市场的各个角落，增强流动性，提升各部位商铺的价值。

建筑形体规整，在保证平面功能布置便捷流畅的前提下，以城市和自然的融合作为设计理念。外立面使用凹凸纹理的要素体现变化的表情，彰显商业氛围。商场入口具有符号性、辨识度，从而吸引人流。大量使用透明感玻璃素材，减少外立面的聚光感。主立面过用两块形似羽翼的折板体现了对自然的融合，既呼应了常熟丰富的自然和历史，又创新了常熟市的地标特色。

总平面

ÆON
ÆON

ÆONMALL

卡滋贝诺

沧浪新城社区服务中心

CANGLANG NEW CITY COMMUNITY SERVICE CENTER

项目类型：社区商业　　**项目地点**：江苏苏州

用地面积：10918m²　　**建筑面积**：27097m²

设计时间：2009年　　**竣工时间**：2012年

第五届全国民营工程设计企业优秀工程设计建筑工程设计类华彩银奖

2015年全国优秀工程勘察设计行业奖建筑工程三等奖

2014年度江苏省第十六届优秀工程设计一等奖

项目位于苏州市沧浪新城，旨在为基地西侧和南侧的居住区提供生活配套服务。设计以“服务为民”和“精神归属”为主题，在满足服务中心多样功能的基础上，给社区居民提供精神归属感。

设计基于沧浪新城的整体城市风貌和 L 形用地现状条件，在紧张的用地上，实现“单一建筑功能”向“建筑和室外场地两种功能兼具”的转变，并达到高效的空间组织模式。错动的建筑形体满足了服务中心多样的功能需求，同时加强了建筑的雕塑感，实现“服务为民”的目的。

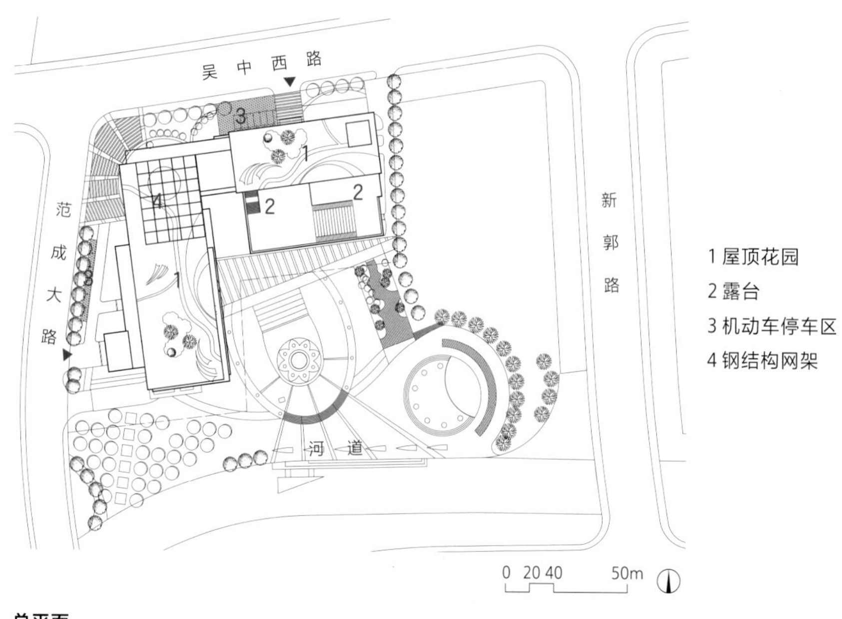

1 屋顶花园
2 露台
3 机动车停车区
4 钢结构网架

总平面

04

文体场馆

CULTURE AND SPORTS

第九届江苏省园艺博览会园博园工程 B 馆
HALL B OF THE 9th JIANGSU HORTICULTURAL EXPOSITION PARK EXPO PROJECT
苏州市妇女儿童活动中心
SUZHOU WOMEN AND CHILDREN'S ACTIVITY CENTER
太湖文化论坛国际会议中心
TAIHU CULTURE FORUM INTERNATIONAL CONFERENCE CENTER
苏州慈济园区
SUZHOU TZU CHI PARK
苏州评弹公园
SUZHOU BALLED SINGING PARK
苏州规划展示馆
SUZHOU PLANNING EXHIBITION HALL
苏州高新区展示馆
SND EXHIBITION HALL
太仓规划展示馆
TAICANG PLANNING EXHIBITION HALL
苏州阳澄国际电竞馆
SUZHOU YANGCHENG E-SPORTS HALL

第九届江苏省园艺博览会园博园工程B馆

HALL B OF THE 9th JIANGSU HORTICULTURAL EXPOSITION PARK EXPO PROJECT

项目类型：文化场馆　　**项目地点**：江苏苏州
用地面积：26047m²　　**建筑面积**：24520m²
设计时间：2013年　　**竣工时间**：2015年
合作单位：直向建筑设计事务所

2017年度全国优秀工程勘察设计行业奖一等奖
2016年度中国建筑学会建筑创作奖银奖
2016年度江苏省第十七届优秀工程设计一等奖
2016年度江苏省城乡建设系统优秀勘察设计一等奖
2019年度江苏省优秀工程勘察设计行业奖暖通二等奖

项目选址于苏州吴中区临湖镇太湖边，承载非物质文化遗产展厅、飞行影院、餐厅等多个功能，在三面临水的生态环境里，钢模清水混凝土餐厅和制高点造型景观塔的建筑造型格外引人注目，是园博园工程的标志性建筑之一。

项目以对自然场地和苏州文化的尊重为整体设计概念的出发点，用院落与连廊连接各个功能空间，营造出园林式的空间体验。考虑苏州当地气候，在多雨的季节，参观者可通过连廊在院落间走动，院落中被置入有故事性的主体建筑，如被绿荫覆盖的球幕影院、标志性的门厅云雾装置、可眺望园区风景的景观塔、面对银杏树林的亲水餐厅等。

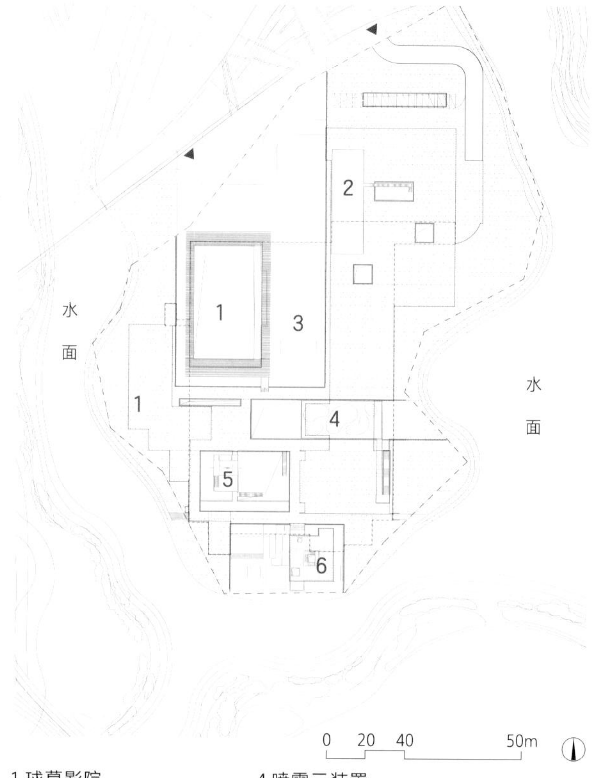

1 球幕影院　　4 喷雾云装置
2 室外展台　　5 景观造型塔 / 下沉庭院
3 室外活动区　　6 亲水餐厅

总平面

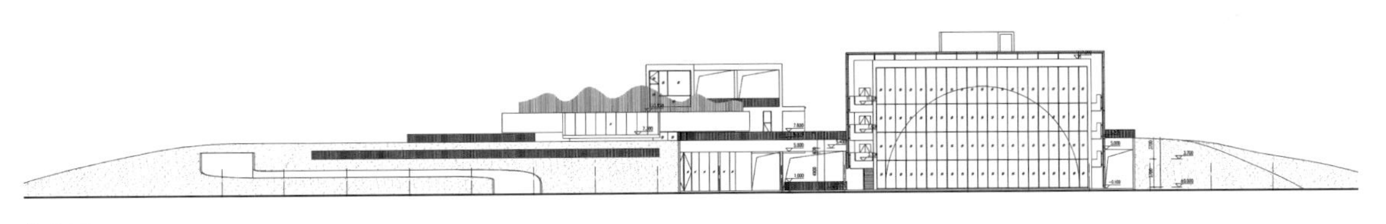

北立面

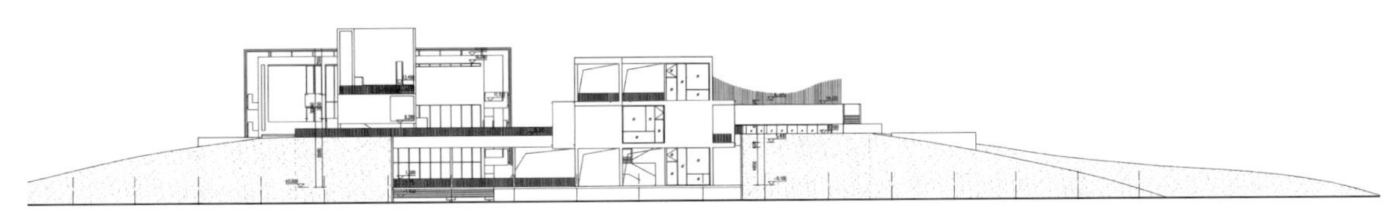

南立面

在功能的需求下，以对自然干扰最小为出发点，利用覆土遮掩住展厅部分建筑体量，进一步强化了建筑与自然的融合。在覆土之下的非物质文化遗产展厅和江南古典园林展厅提供给游人丰富的信息与互动参观体验，三个内庭为展览空间引入自然通风采光，并将游人引向屋顶展览平台；在覆土之上，形成一个拥有丰富植被种类，提供户外表演、就餐、教育体验、亲子互动等功能的公共绿化公园。整个建筑素雅大气、空间丰富、立面简洁，与环境融为一体。

苏州市妇女儿童活动中心

SUZHOU WOMEN AND CHILDREN'S ACTIVITY CENTER

项目类型：活动中心　　**项目地点**：江苏苏州

用地面积：28697m^2　　**建筑面积**：44146m^2

设计时间：2013年　　**竣工时间**：2015年

全国绿色建筑三星级设计标识

2016年度江苏省第十七届优秀工程设计二等奖

2016年度江苏省城乡建设系统优秀勘察设计二等奖

2016年度江苏省优秀工程勘察设计行业奖绿建二等奖

项目位于苏州高新区何山脚下，西邻新区公园，采用“生长的积木”这一概念，用最简单的“方形盒子”来累积建筑的体量，通过各种“盒子”的搭接和错动来适应功能和地形，很多体块以山势、轴线等简单、具象的设计元素“生长”出来。

设计充分利用地下空间、室外透水地面、改善围护结构热工性能、区域雨水收集利用、可调节外遮阳、光导管和下沉庭院采光优化措施、土建与装修一体化等先进技术。通过各项技术的综合利用，项目的整体能耗水平较低，作为以妇女和儿童的培训、教育为主的活动场所，不仅能让孩子看到身边的绿色建筑技术，更可以亲身体会绿色技术带来的益处。

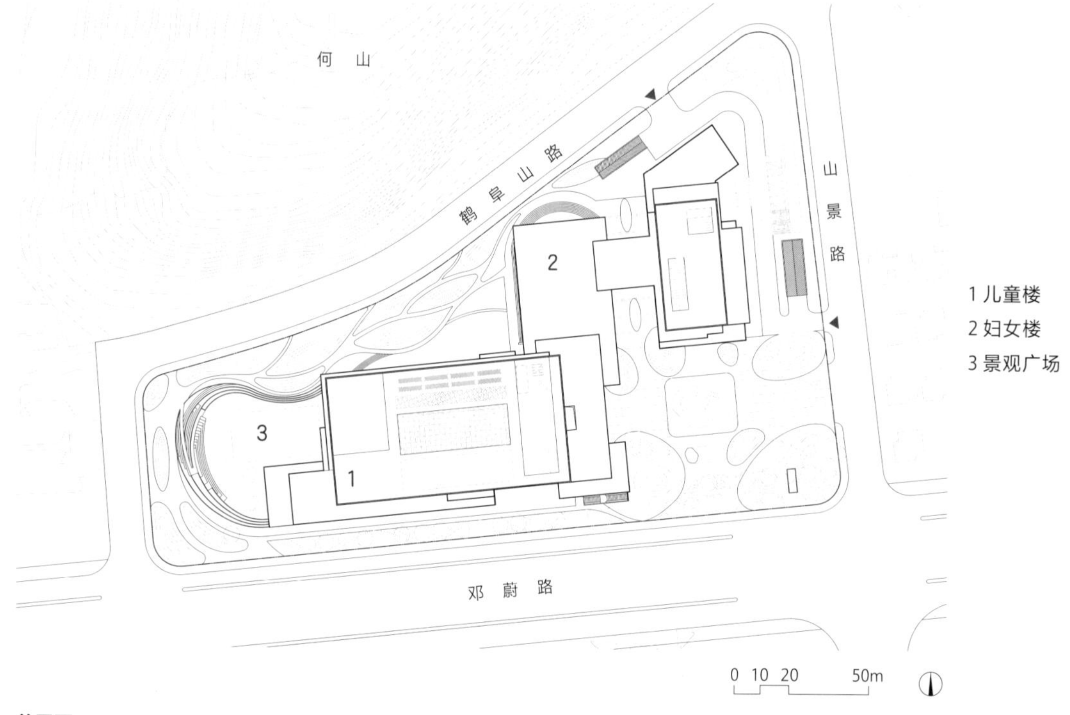

1 儿童楼

2 妇女楼

3 景观广场

总平面

太湖文化论坛国际会议中心

TAIHU CULTURE FORUM INTERNATIONAL CONFERENCE CENTER

项目类型：文化建筑　　**项目地点**：江苏苏州
用地面积：43865m^2　　**建筑面积**：65783m^2
设计时间：2007年　　**竣工时间**：2009年

2012年度江苏省第十五届优秀工程设计二等奖
2011年度江苏省城乡建设系统优秀勘察设计一等奖

项目位于苏州西部风景优美的太湖之滨——香山镇。设计采用散点式的建筑布局，提取了苏州园林的空间尺度和特征元素，将大体量建筑“埋”入山地中，形成层层递进的台地，化解了庞大的形体，为湖面留出了广阔的背景。

根据功能需要，通过钢结构设计将内部空间层层叠加，满足使用需求。在会议中心主体建筑的西南侧，采用散点式的布局方式设计了一组二层高的餐饮接待楼，水面自南侧引入，蜿蜒至餐饮接待楼扩大为一片宁静的池塘，衬托出这一区域的精致、典雅，宛如小家碧玉静立在“姑苏台”之侧。高台与开阔的湖面相映，形成了舒适的视野，同时也与自然山体融合，创造了与环境相一致的天际线。

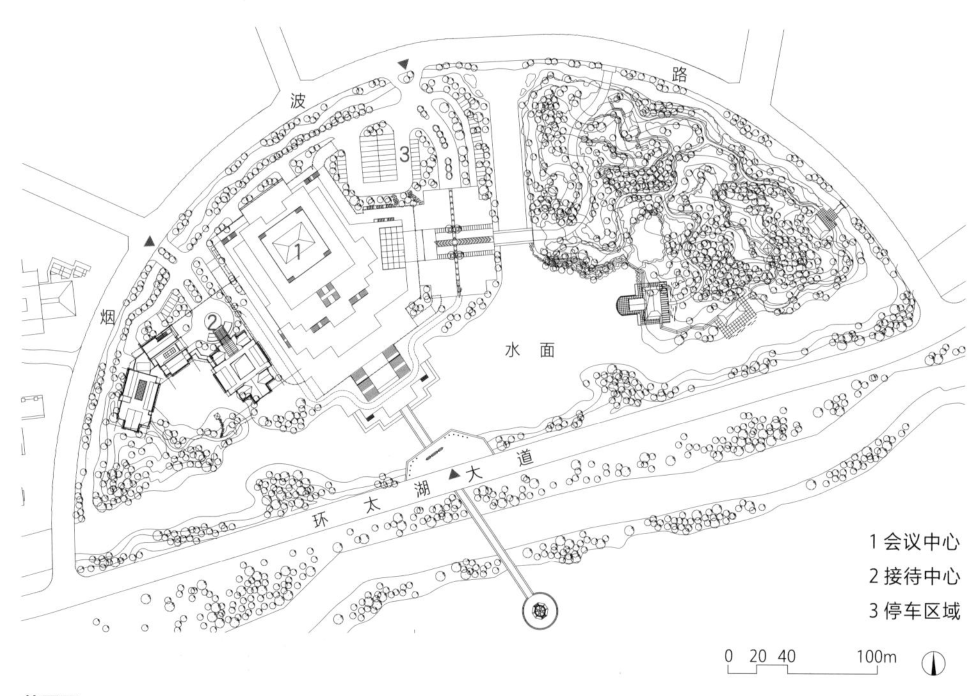

总平面

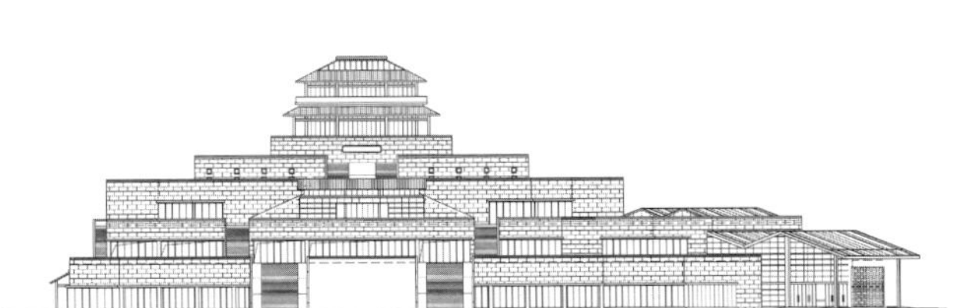

北立面

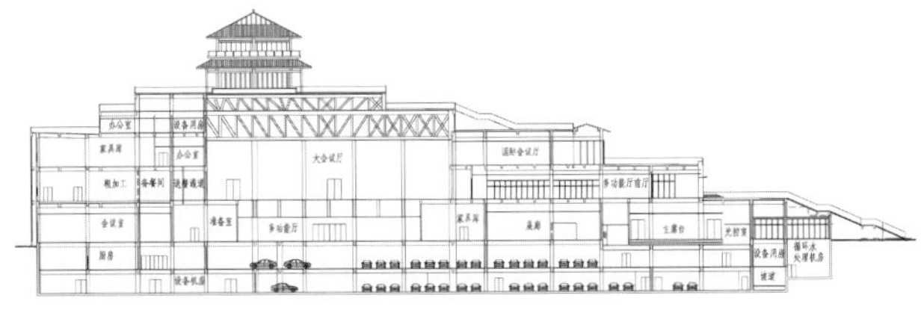

北剖面

苏州慈济园区

SUZHOU TZU CHI PARK

项目类型：文化建筑　　**项目地点**：江苏苏州

用地面积：37261m²　　**建筑面积**：63218m²

设计时间：2006年　　**竣工时间**：2011年

2017年度中勘协传统建筑分会中华建筑文化奖二等奖

2012年度江苏省第十五届优秀工程设计一等奖

2012年度江苏省城乡建设系统优秀勘察设计一等奖

2013年度江苏省优秀勘察设行业奖暖通二等奖

项目北靠景德路，西侧和南侧紧临护城河，是环古城景观带上的一个重要节点。该项目包括静思堂、静思精舍、高科技健检中心、慈济博物馆、慈济培训教育中心和慈济志业中心，是一个功能综合的建筑群。

如何在苏州慈济园区规划与设计中，体现慈济文化体系完整、审美取向鲜明的特色，同时考虑基地所处的特殊位置，探讨苏州城市文脉结构的延展问题，以及建成后的建筑与苏州环古城河景观带的相互影响，是本设计的难点。设计师考虑集约化的土地利用方式，建筑群体化整为零，将建筑群体地下连成一体，地面分栋，化为小体量建筑，在三维维度上组织院落空间，形成集约型院落空间，解决复杂的交通流线，满足现代生活的要求。在满足功能与慈济文化的同时，又在建筑空间、立面造型、色彩细节等方面充分融入苏州传统建筑特色，延续了城市的文脉记忆。整体建筑风格朴实、内敛、雅致，建筑与城市空间融合、共生。

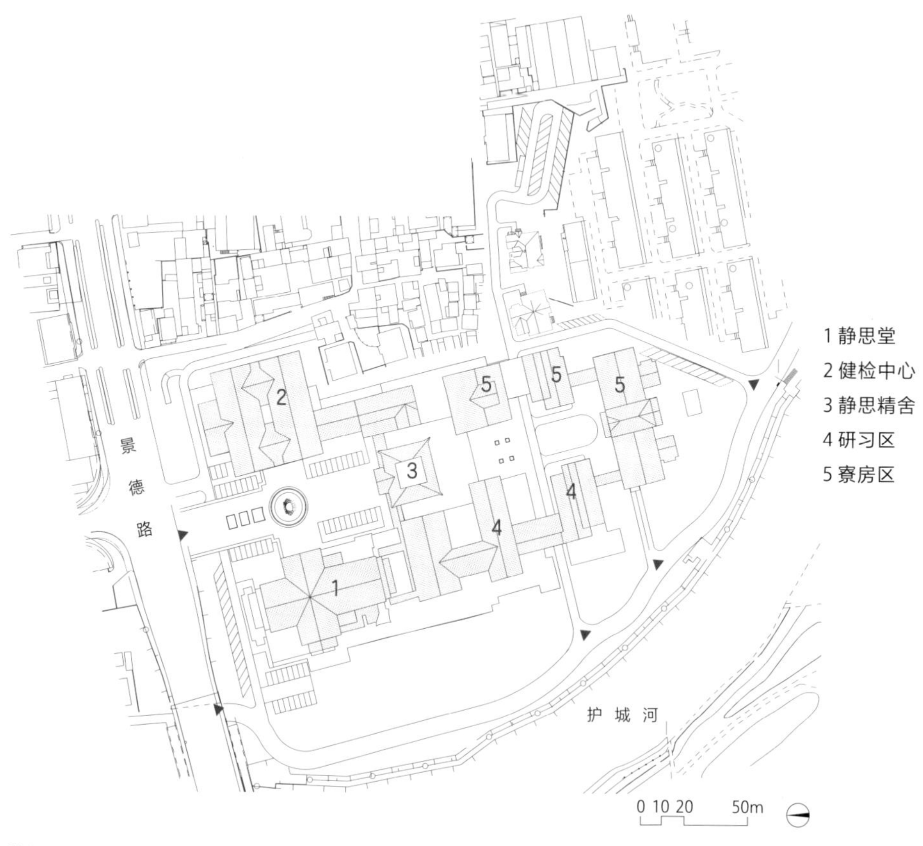

总平面

东立面

西立面

苏州评弹公园

SUZHOU BALLED SINGING PARK

项目类型：文化建筑　　**项目地点**：江苏苏州
用地面积：24333m²　　**建筑面积**：5843m²
设计时间：2020年　　**竣工时间**：2022年
古建合作单位：苏州市计成文物建筑研究设计院有限公司

项目承载的不仅是相城区黄埭书坛老码头的历史记忆，更是苏州评弹的一张名片，居民生活的生动舞台。设计在河边设置的码头成为重要的景观节点，重新演绎过去“江南第一书码头”的热闹景象，在当代城市化的乡镇中，重建一种评弹文化与市民生活的关系。

设计以苏式传统民居粉墙黛瓦为基色，取其进落的基本结构，以山墙间隔划分主次空间，其中围合院落天井，形成丰富的传统意象空间。山墙形式上，采用传统“观音兜”的硬山做法，对其进行现代转译，以连续起伏的山墙演绎评弹宛转悠扬的曲调，沿河一岸，重重民居起伏的屋面，层层叠叠的曲线山墙，与百年评弹的迤逦之音，在场地共鸣。

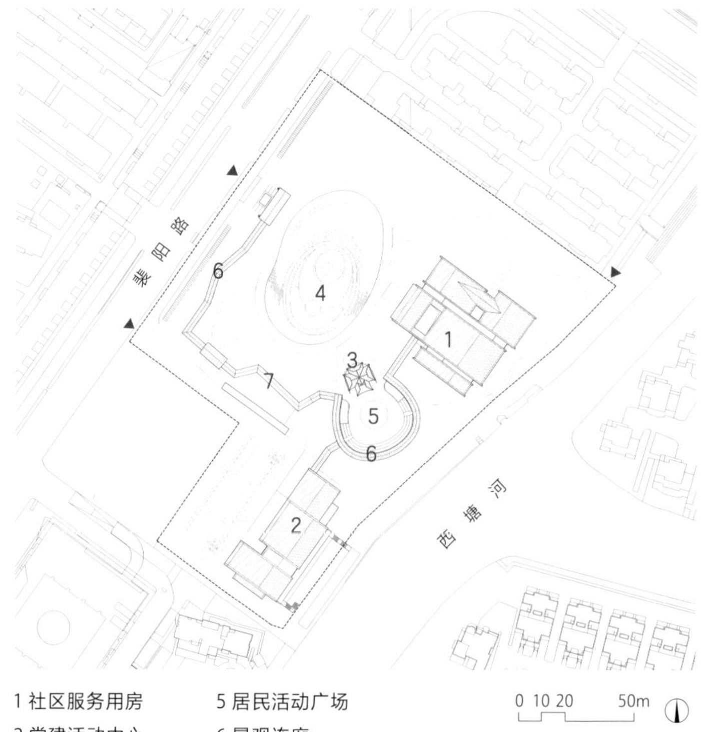

1 社区服务用房
2 党建活动中心
3 古戏台
4 健身场地
5 居民活动广场
6 景观连廊
7 透空景观廊架

总平面

苏州规划展示馆

SUZHOU PLANNING EXHIBITION HALL

项目类型：文化建筑　　**项目地点**：江苏苏州
用地面积：34779m²　　**建筑面积**：19868m²
设计时间：2002年　　**竣工时间**：2003年

2005年度部级优秀勘察设计二等奖
2009年度中国建筑学会建筑创作大奖入围奖

项目总体上分为现代馆、古代馆、万年桥大街两侧辅助用房三部分。北侧的现代馆按照现代展示馆的要求设置空间序列。南侧的古代馆主要由一些保留、移建、新建的民居组成，结合苏式庭院布置，相映成趣。一条自北向南的主轴线结合参观流线，将北部绿化广场、现代馆、中央水广场、古代馆有机地串联起来。建筑采用黑白灰色调，以取得总体环境的协调。

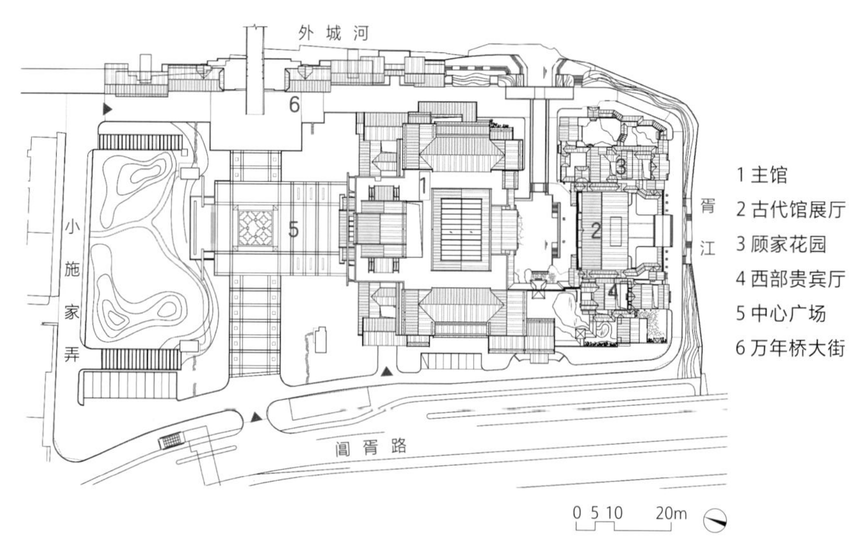

总平面

西立面

东立面

苏州高新区展示馆

SND EXHIBITION HALL

项目类型：文化建筑　　**项目地点**：江苏苏州
用地面积：11069m²　　**建筑面积**：14022m²
设计时间：2011年　　**竣工时间**：2012年
合作单位：百殿建筑设计咨询（上海）有限公司

2014年第五届中国建筑学会优秀暖通空调工程设计奖二等奖
2015年度中勘协民营设计企业分会第五届“华彩奖”银奖
2015年度全国优秀工程勘察设计奖三等奖
2014年度江苏省第十六届优秀工程设计二等奖
2015年度江苏省优秀工程勘察设计绿建二等奖
2013年度省城乡建设系统优秀勘察设计一等奖

项目位于苏州科技城核心区，是科技城智慧谷的门户。项目力求全方位、多视角展示高新区发展、建设及规划的丰硕成果。作为完整体现高新区内涵、风貌与愿景的规划展示馆，主要包含新型城市和多元功能两个主题。设计采用大空间开放立体的展示模式，布局灵活合理。建筑位于基地南部，周边为临时展区，通过绿色河畔景观与人工湖面融合。建筑整体造型寓意“智慧之眼”，审视新区的现在，展望光辉的未来。

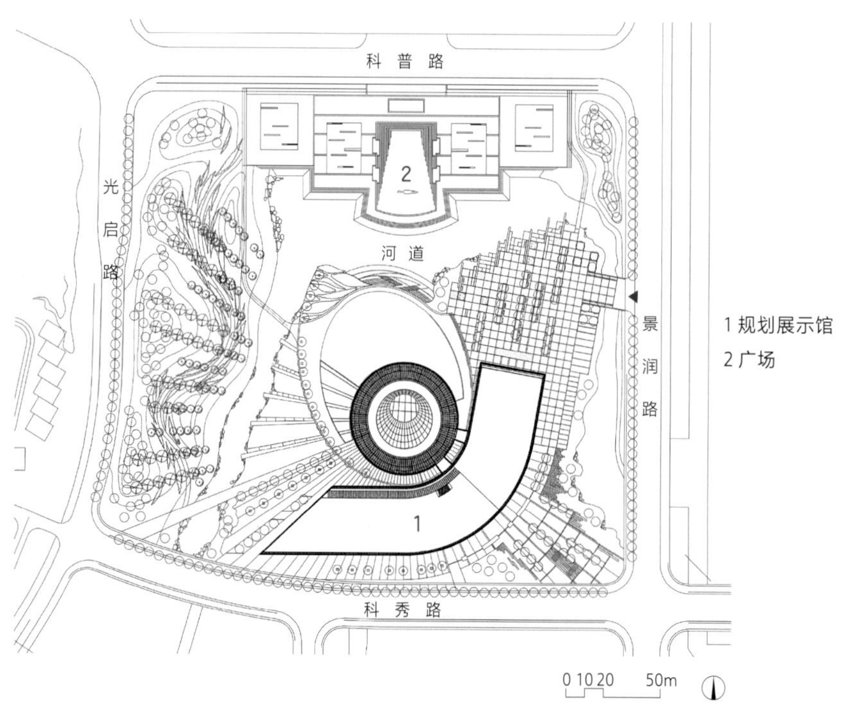

1 规划展示馆
2 广场

总平面

苏州高新区展示馆
SND Exhibition Hall

太仓规划展示馆

TAICANG PLANNING EXHIBITION HALL

项目类型：展示中心　　**项目地点**：江苏太仓
用地面积：20186m²　　**建筑面积**：18545m²
设计时间：2011年　　**竣工时间**：2013年
合作单位：赫睿建筑设计咨询（上海）有限公司

项目位于太仓市科教新城天镜湖畔，由延伸于湖面的亲水平台及一边位于平台上、一边漂浮于水上的主体展览建筑组成，是集展示、宣传、交流等功能于一体的专业性展览馆，也是太仓市第一座雕塑型建筑，现已成为展示太仓新形象的重要平台、对外宣传和招商引资的重要窗口、公众参与和交流城市规划建设的重要场所。

设计以“双鱼”为灵感，主体展馆造型仿佛两条相互追逐、欢快游动的锦鲤，集灵动创新于一体。项目空间造型复杂，鱼形单层网壳采用近万块尺寸各异的金属板，屋面由18块标高不同的三角形玻璃组成，玻璃顶棚跟鱼身为曲面相交，立面幕墙采用了大量双曲面玻璃。设计时首次尝试BIM辅助设计，以空间模型定位结构构件，并进行结构计算分析与模拟，很好地满足了设计要求，实现了实用性与经济性的统一。

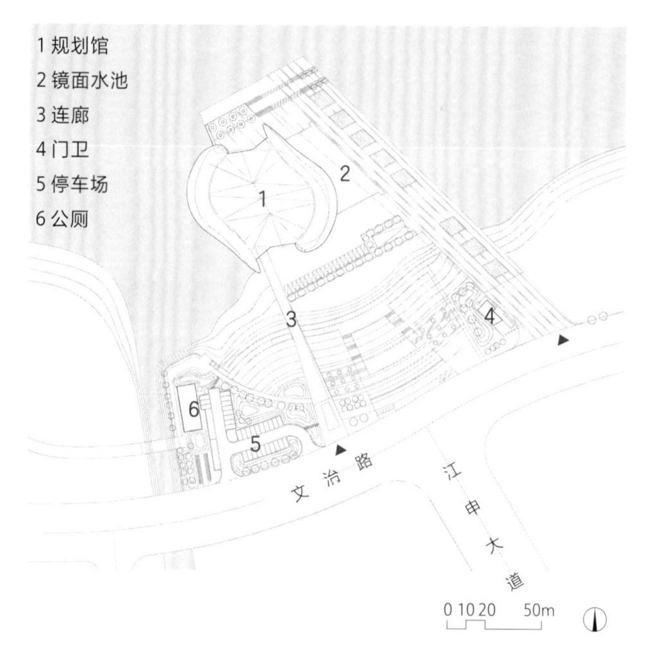

总平面

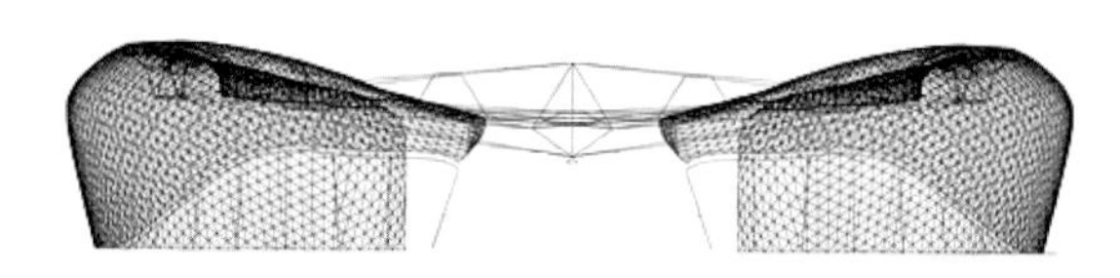

钢结构立面

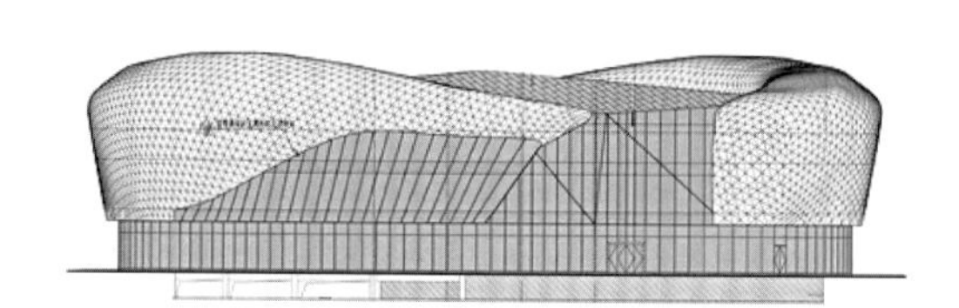

南立面

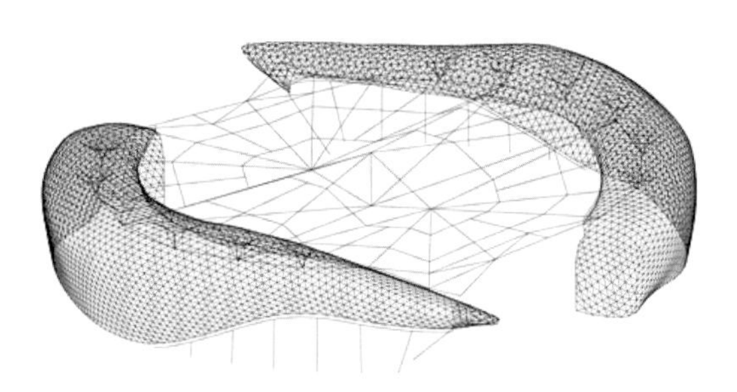

钢结构三维

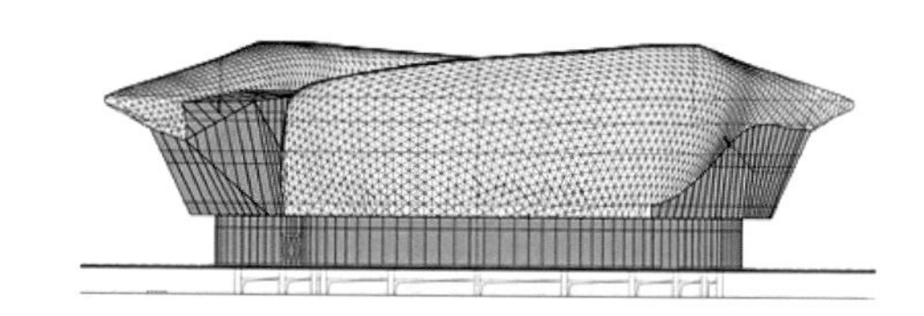

东立面

苏州阳澄国际电竞馆

SUZHOU YANGCHENG E-SPORTS HALL

项目类型：会展中心　　**项目地点**：江苏苏州

用地面积：21828m^2　　**建筑面积**：9542m^2

设计时间：2018年　　**竣工时间**：2019年

2020年第江苏省十九届优秀工程设计三等奖

2020年江苏省城乡建设系统优秀勘察设计二等奖

2018年度江苏省土木建筑学会第十二届建筑创作奖二等奖

项目位于高铁新城环秀湖边，临近苏州高铁北站。项目作为新型体育建筑，聚合电竞赛事和周边内容制作，打造集“产业、文化、教育”于一体的综合性电竞产业生态圈和场馆品牌。

如何满足电竞行业的先锋、竞技特点，同时融入高铁新城快速发展的城市风貌之中，成为环秀湖畔、高铁线旁的标志建筑，是本项目的难点。设计在布局上结合景观向心布置，建筑体量在场地中以巨石形象坐落其中，形态犹如环秀湖边散落的石块，外表银灰色铝板的曲线扭转与不远处的苏州高铁北站、高铁线相呼应，为高铁新城增添了一份动感。

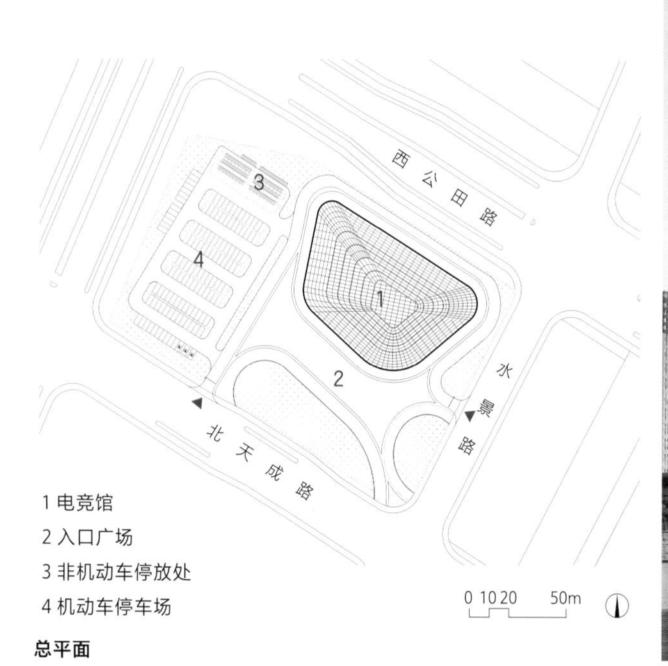

总平面

05

酒店建筑
HOTEL BUILDING

苏州金普顿竹辉酒店
KIMPTON BAMBOO GROVE CHIC HOTEL IN SUZHOU
苏州太湖万丽万豪酒店
SUZHOU TAIHU WANLI MARRIOTT HOTEL
东山宾馆
DONGSHAN HOTEL
裸心泊酒店
NAKED WATER

苏州金普顿竹辉酒店

KIMPTON BAMBOO GROVE CHIC HOTEL IN SUZHOU

项目类型：酒店建筑　　**项目地点**：江苏苏州

用地面积：43000m^2　　**建筑面积**：61000m^2

设计时间：2016年　　**竣工时间**：2022年

合作单位：水石设计

项目位于苏州老城区姑苏区内，南临竹辉路，北临南园河，东侧为南园河支流，西侧和西北侧为南石皮弄。基地位置文化底蕴丰厚，地理条件得天独厚，周边历史人文旅游资源丰富，交通便捷。

旧竹辉饭店作为姑苏地标性建筑，代表着高端和精致的苏州特色，是苏州的时代印记。如何将“传承”与“创新”相融合，在重塑竹辉记忆的同时注入新的时代特色，成为本项日的最人挑战。设计将北侧规划为酒店片区，南侧设置为酒店配套餐饮片区。场地中部生成轴线，联动南北，贯通东西。以东西横轴为水轴，延续老竹辉和苏州园林以水为隔、隔而不离的布局方式，作为新竹辉商业和酒店的相接界面。设计着重打造苏式园林空间，园内有园，园外有景，以水为主，主题突出。同时结合空间布局自然生长出竹辉八景，串联起整个园区建筑：行馆－朗轩－雅集－戏韵－樟庭－亭湖－竹辉，亭台水榭相映生辉。另外酒店区域也延续了老竹辉饭店临湖而筑的构思，建筑景观倒影水中，闪烁出无限意境。景观设计保留老竹辉饭店太湖石，植入新竹辉饭店景观庭院内，在翠竹掩映之中，让历史悄然传承。同时建筑单体延续了苏式建筑的粉墙黛瓦，提炼古建筑元素的双坡顶，并加上现代的处理手法，使园区与周边的古城建筑融为一体，更加契合。

1 金普顿酒店大堂
2 酒店客房区
3 宴会厅
4 环宇荟步行街
5 大堂吧
6 水景
7 竹辉商厦（已建）

总平面

苏州太湖万丽万豪酒店

SUZHOU TAIHU WANLI MARRIOTT HOTEL

项目类型：酒店建筑　　**项目地点**：江苏苏州

用地面积：92450m²　　**建筑面积**：118408m²

设计时间：2013年　　**竣工时间**：2018年

合作单位：意境（上海）建筑设计有限公司

2015年度第三届全国BIM大赛优秀奖

2019年度省优秀工程勘察设计行业奖电气三等奖

2020年度省第十九届优秀工程设计三等奖

2019年度省城乡建设系统优秀勘察设计三等奖

2014年度江苏省建筑信息模型（BIM）应用设计大赛最佳BIM建筑设计奖、最佳BIM工程协同奖

项目位于苏州吴中区胥口镇渔洋山东南侧环太湖区域，东临墅里路，西临舟山路，南临太湖，北依渔阳山群，交通便利，地势平坦。地块内包含万豪和万丽两个酒店。

本项目设计灵感源于太湖的诗情画意，通过引入香山匠人的工艺，使苏州园林的韵味与场地环境和谐共生。设计统筹考虑用地、景观、朝向、通风以及经营模式等一系列因素，充分结合万丽、万豪品牌的经营理念，在视觉和空间上拉开两个酒店的距离，营造出完全不同的两种体验。万丽酒店与山呼应，以横向线条为主，建筑整体明亮舒展；万豪酒店与太湖呼应，以竖向线条为主，采用灰色石材装饰，整个建筑硬朗挺拔，凸显商务氛围。

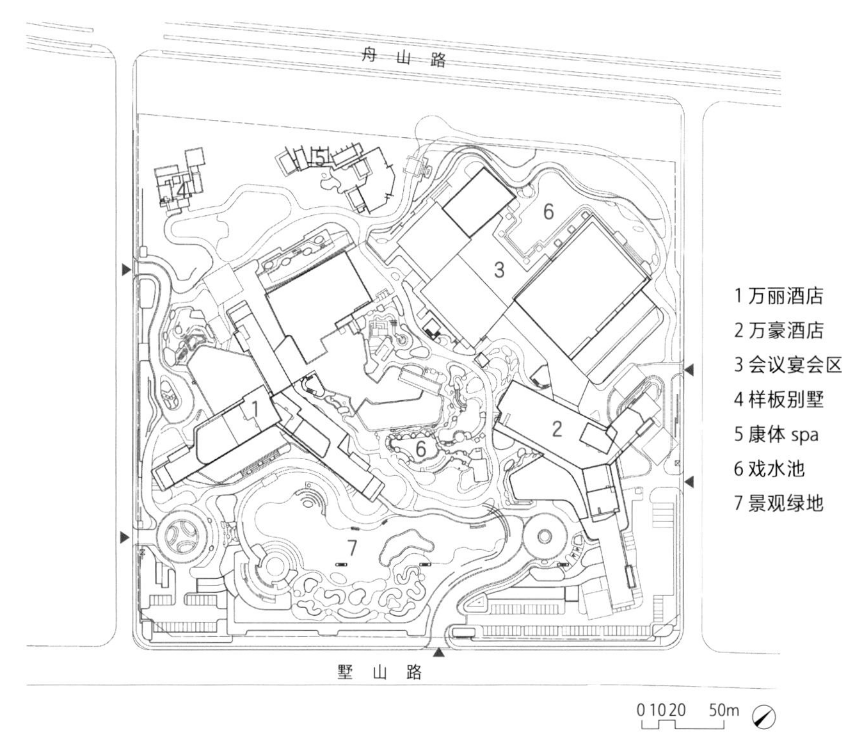

总平面

东山宾馆

DONGSHAN HOTEL

项目类型：酒店建筑　　**项目地点**：江苏苏州
用地面积：103337m²　　**建筑面积**：42293m²
设计时间：2004年　　**竣工时间**：2016年

2008年度国家优质工程银质奖
2017年度全国优秀工程勘察设计行业奖之“华筑奖”三等奖
2017年度中勘协传统建筑分会中华建筑文化奖三等奖
2012年度江苏省第十五届优秀工程设计二等奖
2012年度江苏省城乡建设系统优秀勘察设计二等奖

东山宾馆三期综合楼东临太湖，南面与历史悠久的苏州园林启园毗邻，是接待国家级首长及外国元首的基地。项目以苏州传统文化为根本，结合现代建构技术，尽显时代精神和苏州特色。

如何满足国宾馆的特点，同时融入周边的传统文化风貌之中，打造苏式园林建筑，是本项目的难点。设计摈弃了豪华、奢靡的负担感，多了几分山水环绕的家园感。说是国宾馆，感觉又如私家园林，既庄重又亲切。为减少西面环山公路、民房对宾馆的干扰，西侧运用了人工堆山植树的阻挡方法，并将三期的动力中心藏于土丘之下。而面向太湖的东侧则完全打开，美妙湖景尽收眼底。以自然景观为主导的东山宾馆三期工程，将建筑尽可能掩映在山水绿化之间；利用独特的地貌环境，体现建筑的坡地特征；建筑造型与周围环境互相协调，使其在自然环境中体现平和、朴实的风格。

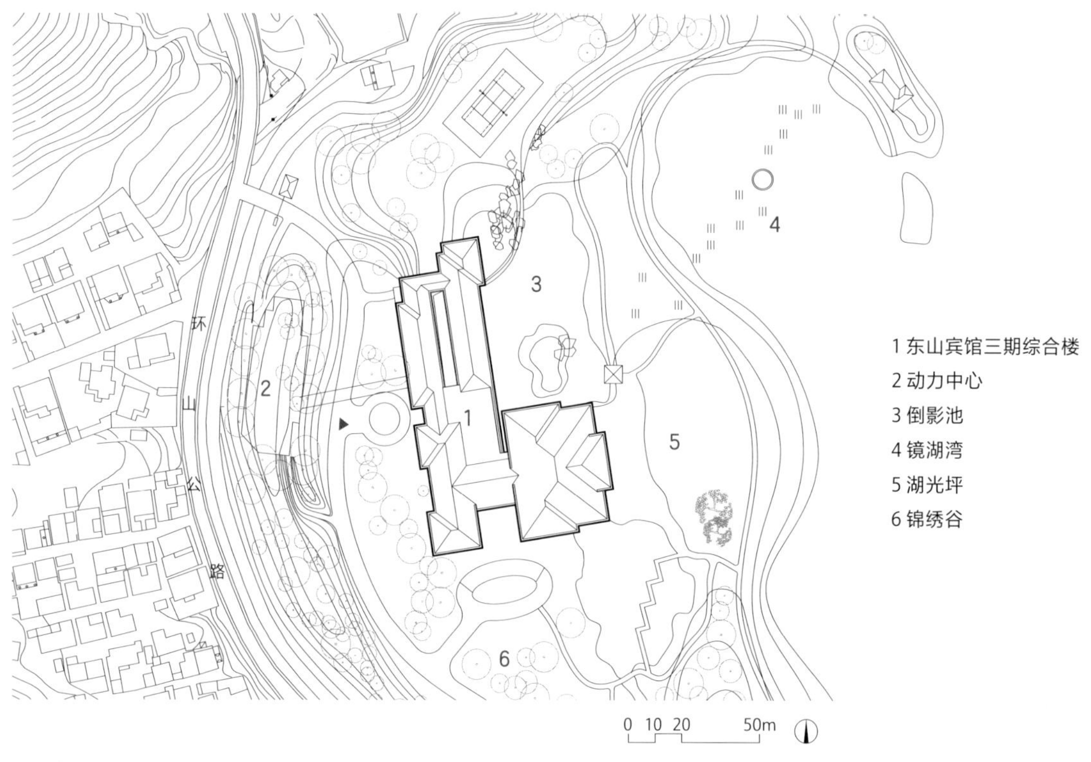

东山宾馆三期综合楼总平面

东山宾馆总统楼位于山顶，建筑本身也成为了湖景的一部分。通过对建筑的弱化处理，使太湖景观呈现自然本色，并能和其他建筑协调。受原建筑的限制，建筑平面呈三角形和长方形的组合，在建筑中部设计了一个圆形的小中庭，化解了几何形体拼合产生的矛盾，创造出一个富有特色的空间。

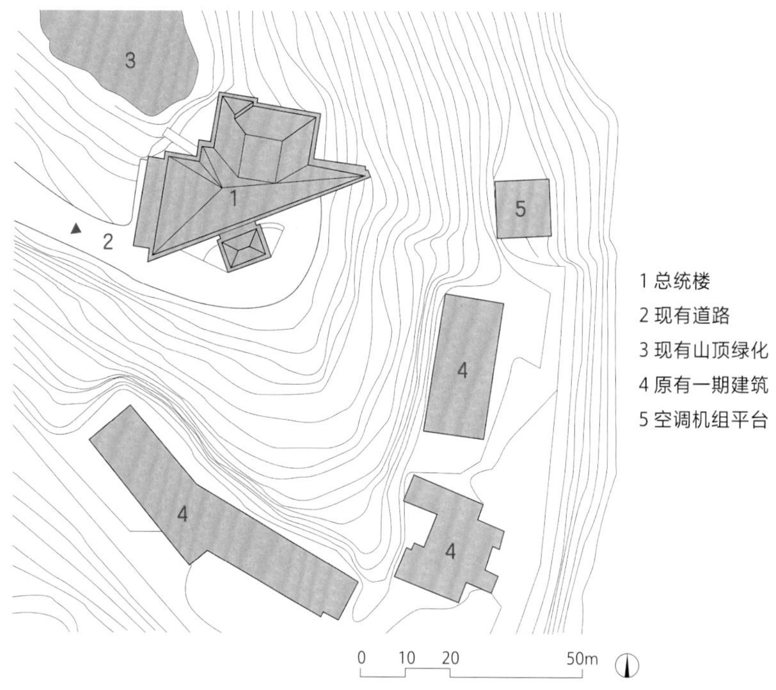

1 总统楼
2 现有道路
3 现有山顶绿化
4 原有一期建筑
5 空调机组平台

东山宾馆总统楼总平面

东山宾馆叠翠楼与东山宾馆整体风格相协调，延续了周边建筑庄重、大气的特点，总体显得低调、内敛。建筑体量后高前低，依山就势。屋面采用平坡结合，层层退台，稳重而不失活力。设计上“一减一增”，通过削减宾馆东部入口处的建筑体量，增加一个观景平台，太湖和启园的美景尽收眼底。

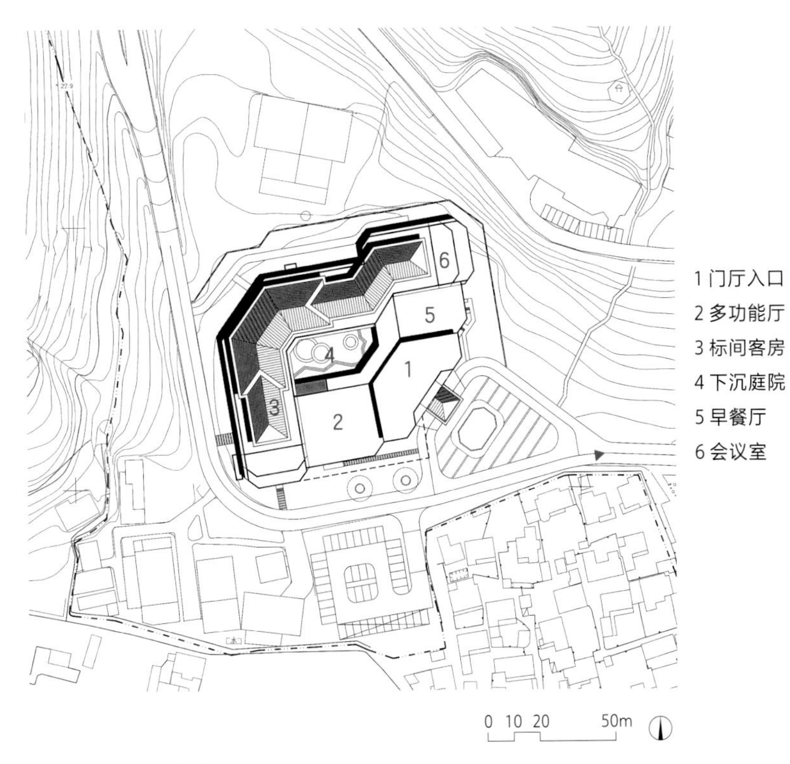

东山宾馆叠翠楼总平面

裸心泊酒店
NAKED WATER

项目类型：酒店建筑　　**项目地点**：江苏苏州
用地面积：32360m²　　**建筑面积**：13760m²
设计时间：2013年　　**竣工时间**：2019年
合作单位：裸心酒店管理（上海）有限公司

2021年度江苏省城乡建设系统优秀勘察设计二等奖
2022年度江苏省第二十届优秀工程设计三等奖

项目坐落于太湖之畔的镇湖。基于“虽由人作，宛自天开”的设计理念，营造“功能完善、亲近自然、具有地域特色、可持续发展”的顶级亲水度假酒店区。在建造过程中最大程度地保留和利用自然地形，融合当地文化的精髓，采用当地材料，通过不同材质的重新组合、颜色之间的拼接演绎，创造出令人耳目一新又充满野趣的度假休闲场所。酒店整体采用散落式布局，局部公共区域为院落组团式布局，建筑和景观相互融合，移步异景，仿佛置身园林，营造出曲折清幽的游园体验。

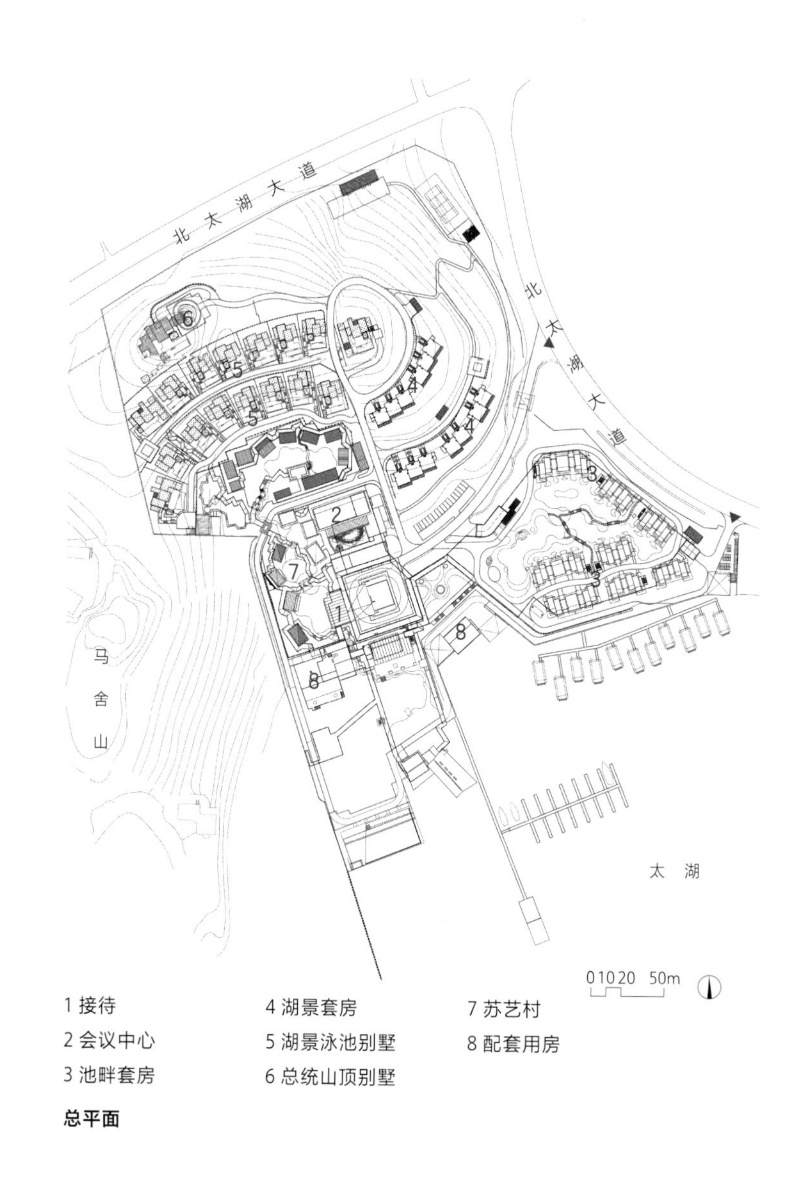

1 接待　　4 湖景套房　　7 苏艺村
2 会议中心　　5 湖景泳池别墅　　8 配套用房
3 池畔套房　　6 总统山顶别墅

总平面

naked

06

教育建筑

EDUCATIONAL ARCHITECTURE

苏州大学理工实验楼
SCIENCE AND ENGINEERING EXPERIMENT BUILDING OF SOOCHOW UNIVERSITY
中国科学技术大学苏州研究院仁爱路校区
REN'AI ROAD CAMPUS OF SUZHOU RESEARCH INSTITUTE OF UNIVERSITY OF SCIENCE AND TECHNOLOGY OF CHINA
中国中医科学院大学
UNIVERSITY OF CHINESE ACADEMY OF TRADITIONAL CHINESE MEDICINE
南通大学啬园校区
NANTONG UNIVERSITY SEYUAN CAMPUS
西交利物浦大学行政信息楼
XI'AN JIAOTONG-LIVERPOOL UNIVERSITY ADMINISTRATION INFORMATION BUILDING
西交利物浦大学科研楼
XI'AN JIAOTONG-LIVERPOOL UNIVERSITY SCIENTIFIC RESEARCH BUILDING
中国常熟世界联合学院
UWC CHANGSHU CHINA
江苏省木渎高级中学
JIANGSU PROVINCE MUDU SENIOR HIGH SCHOOL
江苏省苏州实验中学科技城校
JIANGSU SUZHOU EXPERIMENTAL MIDDLE SCHOOL SCIENCE AND TECHNOLOGY CITY CAMPUS
苏苑高级中学
SUYUAN SENIOR HIGH SCHOOL
吴江中学初中部
MIDDLE SCHOOL DEPARTMENT OF WUJIANG HIGH SCHOOL
苏州高新区实验初中东校区扩建
SND EXPERIMENTAL MIDDLE SCHOOL EAST CAMPUS EXPANSION
梁丰初级中学西校区
WEST CAMPUS OF LIANGFENG JUNIOR HIGH SCHOOL
星海实验中学
XINGHAI EXPERIMENTAL MIDDLE SCHOOL
常熟市三环小学
CHANGSHU SANHUAN PRIMARY SCHOOL
吴郡幼儿园
WUJUN KINDERGARTEN
重庆哈罗国际学校
HARROW CHONGQING
伊顿国际学校
ETON HOUSE INTERNATIONAL SCHOOL

苏州大学理工实验楼

SCIENCE AND ENGINEERING EXPERIMENT BUILDING OF SOOCHOW UNIVERSITY

项目类型：教育建筑　　**项目地点**：江苏苏州
用地面积：7200m²　　**建筑面积**：25200m²
设计时间：1999年　　**竣工时间**：2001年

2002年度江苏省第十届优秀工程设计奖一等奖
2001年度江苏省城乡建设系统优秀勘察设计一等奖

苏州大学是一所近百年历史的著名学府，“双塔为笔，方塔为墨，荷花池为砚”成为创作的源泉，设计构思时作了以下考虑：

借景：充分利用“方塔”这一苏州大学的历史标志，新建大楼中轴线与校园南北主轴线重合，中轴线底部五层架空，形成一个24m×16.5m的巨大门洞，使方塔一直处于南北轴线视线范围内，此为“借景”。

造景：在北校门与理工实验楼之间的主广场上，沿中轴线布置一系列水池，内植荷花，寓意“重现荷花池”，广场中间设置一个象征苏大科技腾飞的雕塑，此为“造景”。

移景：门洞两侧为优美的双塔造型，与历史文化建筑“双塔”产生对话，通过巧妙地移植这一意向，使双塔和方塔在同一空间内相映成趣，此为“移景”。

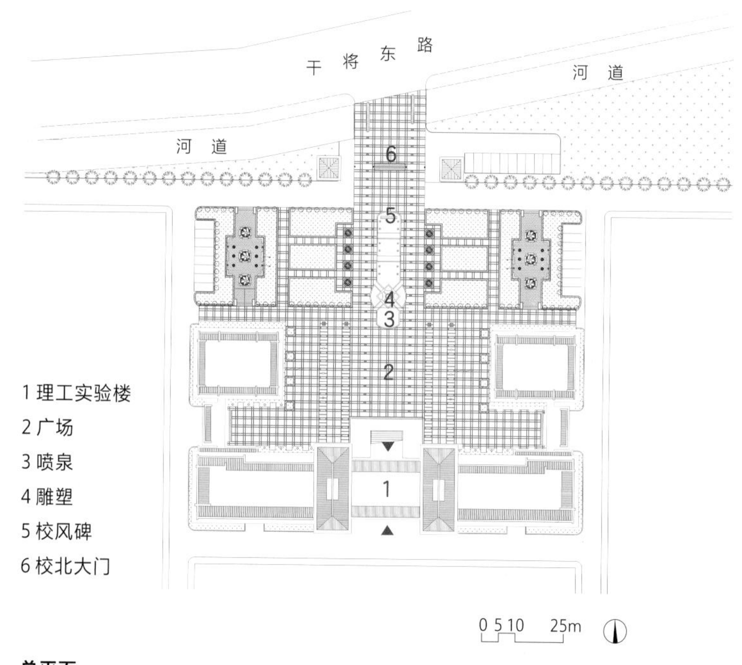

总平面

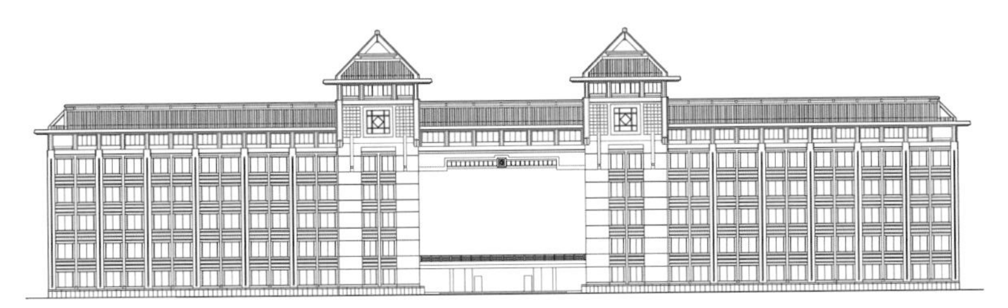

南立面

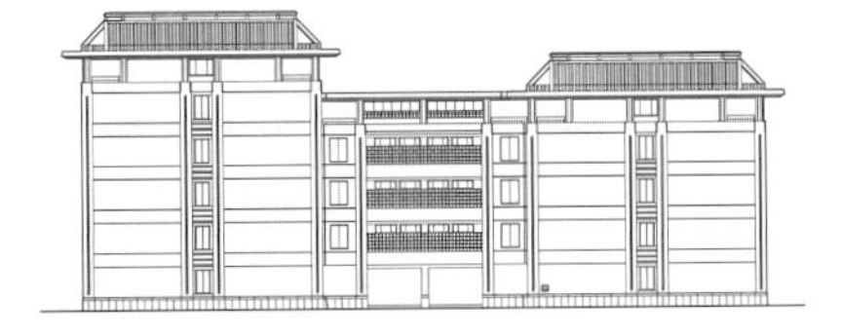

东立面

中国科学技术大学苏州研究院仁爱路校区

REN'AI ROAD CAMPUS OF SUZHOU RESEARCH INSTITUTE OF UNIVERSITY OF SCIENCE AND TECHNOLOGY OF CHINA

项目类型：教育建筑　　**项目地点**：江苏苏州
用地面积：20200m²　　**建筑面积**：146000m²
设计时间：2021年　　**竣工时间**：在建

2021年度江苏省土木建筑学会第十五届“建筑创作奖”一等奖

项目位于苏州工业园区，为传承中科大办学特色与文化内涵，形成功能实用、布局灵活、具有视觉冲击的中科大特色建筑风格的百年校园。设计采用红砖元素、红砖肌理提升整个建筑环境的学术氛围。

在空间布局上，科研中心以柔性的边界促成与周边城市功能的良好融合，通过功能与外在建筑形式的双重尝试，打造校城一体的开放式共享系统，以校区内的弧形架空天桥作为“超级纽带”，从校园西北侧的斜塘庙与永安桥开始，串联起孺子牛广场、再西湖、问渠亭、梅花广场，再到教师公寓生活广场，连接场所记忆。校园的核心位置是生态纽带的主体，河两侧的新建建筑与已有建筑环抱围合形成核心景观区，通过打造二层步行环道系统，使各不同功能空间在二层无缝对接，增强连接性的同时，提升了整个校园的向心性。

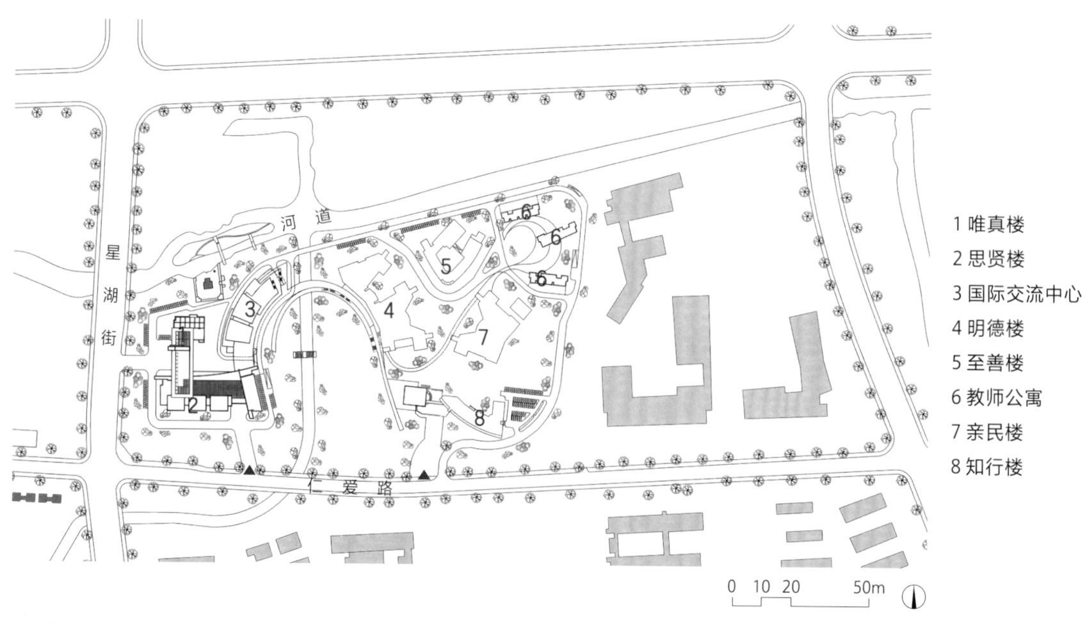

总平面

中国中医科学院大学

UNIVERSITY OF CHINESE ACADEMY OF TRADITIONAL CHINESE MEDICINE

项目类型：教育建筑　　**项目地点**：江苏苏州

用地面积：406677m²　　**建筑面积**：321660m²

设计时间：2022年　　**竣工时间**：在建

合作单位：清华大学建筑设计研究院有限公司

项目位于苏州市吴中区临湖镇园博园东北，共设有32栋主要建筑。建筑设计力求体现以人为本、功能完善、生态和谐，既体现传统文化，又具有时代感。校园设施包含教学楼、实验楼、图书馆、行政办公楼、宿舍楼、食堂、会议中心、体育馆等。景观规划充分尊重场地原有水文条件，在现状河道上稍作调整，在校园中心区将水域扩大为湖面，形成校园景观核心，各类公共建筑及景观节点围绕湖边布局。整个校园以水系为脉络，形成动静分明、层次丰富的园林式景观体系。

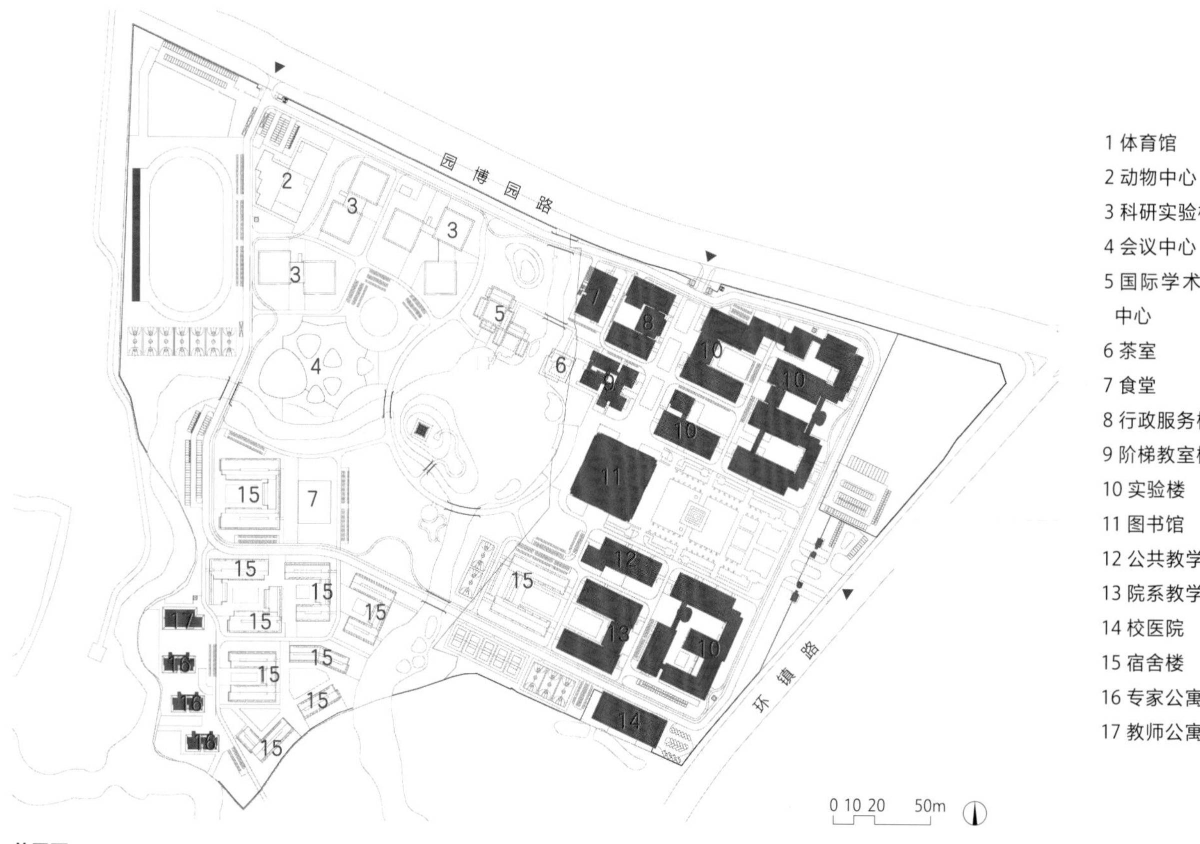

总平面

南通大学啬园校区

NANTONG UNIVERSITY SEYUAN CAMPUS

项目类型：教育建筑　　**项目地点**：江苏南通
用地面积：1271830m²　　**建筑面积**：1069530m²
设计时间：2020年　　**竣工时间**：在建
合作单位：清华大学建筑设计研究院有限公司

设计以南通江海文化为底蕴，以水系为校园之脉，传承南通大学百年历史，设计延续原有校区的空间布局模式、尺度肌理及建筑风格。在保留既有生态本底的基础上，通过多层次公共空间和慢行系统进行有机组织，打造疏密有致、富有呼吸感的花园生态环境。设计力求体现信息时代大学校园建筑特色，结合新型教学方式设计灵活开放的教学空间，激发高等院校的创新性与创造力。

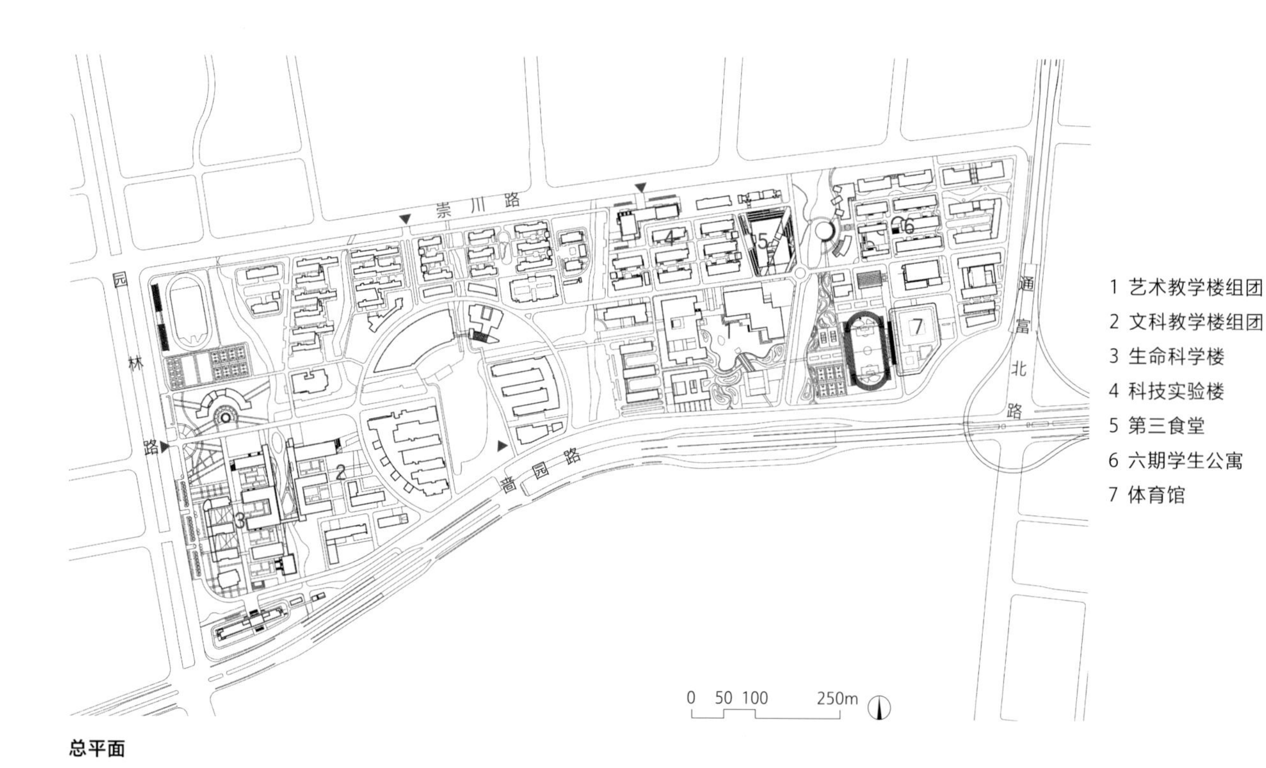

总平面

COLLEGE OF
LIFE SCIENCE

西交利物浦大学行政信息楼

XI'AN JIAOTONG-LIVERPOOL UNIVERSITY ADMINISTRATION INFORMATION BUILDING

项目类型：教育建筑　　**项目地点**：江苏苏州

用地面积：156597m²　　**建筑面积**：59922m²

设计时间：2009年　　**竣工时间**：2013年

合作单位：凯达环球建筑设计咨询（北京）有限公司

2015年度全国优秀工程勘察设计行业奖建筑工程二等奖

2015年度第九届全国优秀建筑结构设计奖二等奖

2016年度中国建筑学会建筑创作奖入围奖

2015年度江苏省城乡建设系统优秀勘察设计一等奖

2015年度江苏省第十七届优秀工程设计二等奖

2015年度两岸四地建筑设计专业组商业办公楼类金奖

2014年度江苏省优秀工程勘察设计行业奖建筑结构二等奖

该项目作为西交利物浦大学北校区的中心建筑，既是学校行政管理和公共活动的重要场所，也是全校师生开展多层次交流的主要场所。

设计概念源于"太湖石"，既是对建筑所处地域文化的呼应，也通过内外连通的孔洞使建筑更好地融入于校园环境中。两层高的裙房从北广场以绿坡缓缓升起，和屋顶花园连为一体，建筑仿佛从地下"破土而出"。塔楼的造型由"太湖石"切削而来，外轮廓呈立方体，以呼应北校区井然有序的建筑肌理；内部的孔洞作为各功能区之间的空间分隔，自然地形成了贯穿多层的空中花园。外立面的开洞使建筑成为一个可呼吸的有机体，也提高了这座校园中心建筑的吸引力。

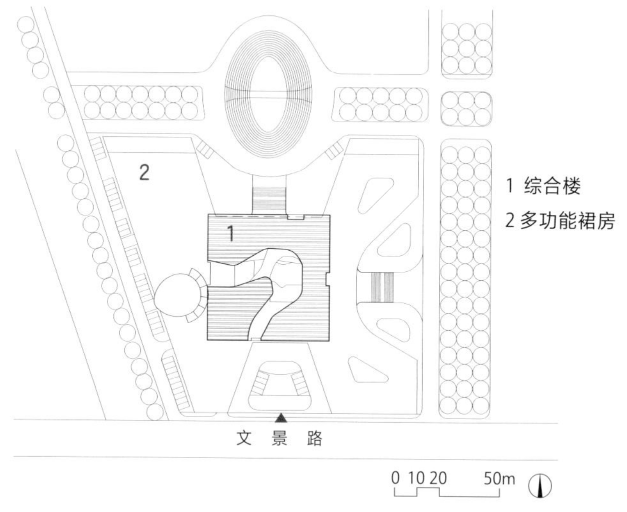

总平面

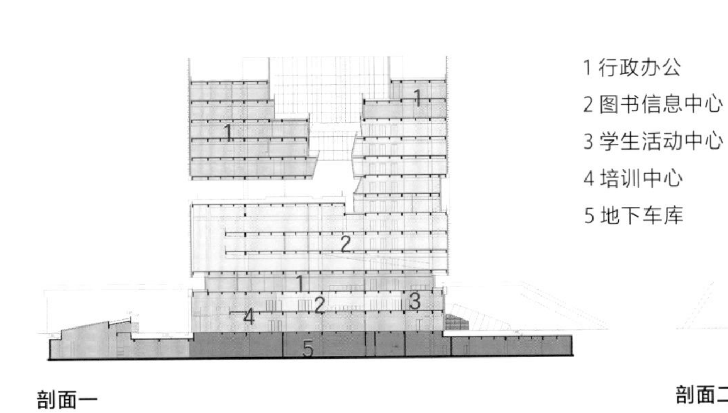

剖面一

1 行政办公
2 图书信息中心
3 学生活动中心
4 培训中心
5 地下车库

剖面二

西交利物浦大学科研楼

XI'AN JIAOTONG-LIVERPOOL UNIVERSITY SCIENTIFIC RESEARCH BUILDING

项目类型：教育建筑　　**项目地点**：江苏苏州

用地面积：34363m²　　**建筑面积**：45041m²

设计时间：2008年　　**竣工时间**：2010年

合作单位：美国帕金斯威尔设计事务所

2013年度全国优秀工程勘察设计行业奖一等奖

2015年度第九届全国优秀建筑结构设计奖结构三等奖

2012年度江苏省第十五届优秀工程设计一等奖

2011年度江苏省城乡建设系统优秀勘察设计一等奖

2011年度江苏省优秀工程勘察设计行业奖结构二等奖

项目位于苏州工业园区独墅湖高等教育区内，科研楼是西交利物浦大学北校区的重点建筑，设计上通过合理的规划构思和先进的绿色技术，为师生提供开放、舒适的办公和学习环境。建筑通过自身围合形成若干个广场，作为学生的室外交流空间。建筑的西面为绿化用地，既是校区绿化广场的延续，也是科研楼自身小环境的创造。南北立面采用大片的点式玻璃幕墙，和东西立面的大片实墙构成强烈的虚实对比。

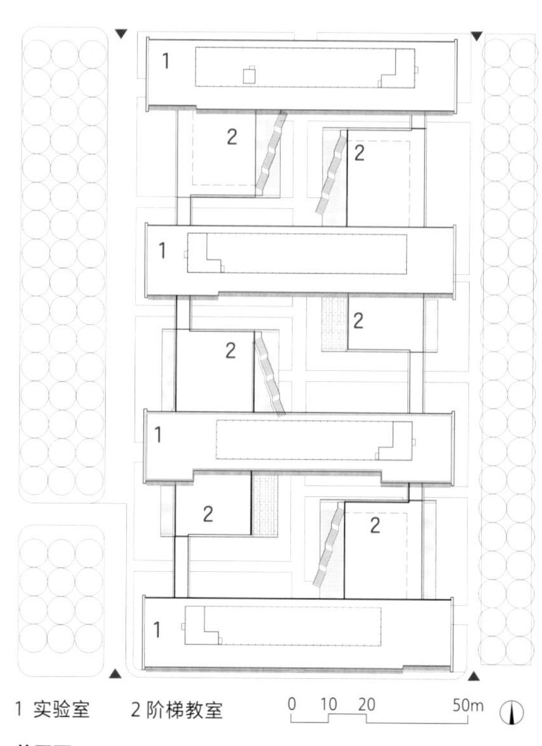

总平面

中国常熟世界联合学院

UWC CHANGSHU CHINA

项目类型：教育建筑　　**项目地点**：江苏苏州
用地面积：85000m²　　**建筑面积**：79035m²
设计时间：2012年　　**竣工时间**：2018年
合作单位：莫平建筑设计顾问(北京)有限公司

项目位于常熟市，东、南、北侧与昆承湖相接。书院内主要包括百家坛、礼堂和戏剧音乐中心、运动健身中心、招待中心、国学书院等建筑。

设计力求在建筑细节上体现出UWC的风格与教育理念。在总体规划布局上，学校以“百家坛”为中心点，将学校各个功能区域分布在四面八方，用几何学突出几何美，从迈入学校大门到学校任意区域都能感受到置身于花园式现代学府的气息。

校园设计围绕着学校环境优化和可持续发展教育的实施，因地制宜地将绿色、节能、生态、舒适的校园环境和教学实践融合。采用能耗监测、能效监测、碳排监测、三维可视化BIM运维平台、窗磁人体感应、数据展示与发布系统等，为校园的绿色运营提供了便捷高效的智能信息工具，营造出舒适自然的国际生态校园。

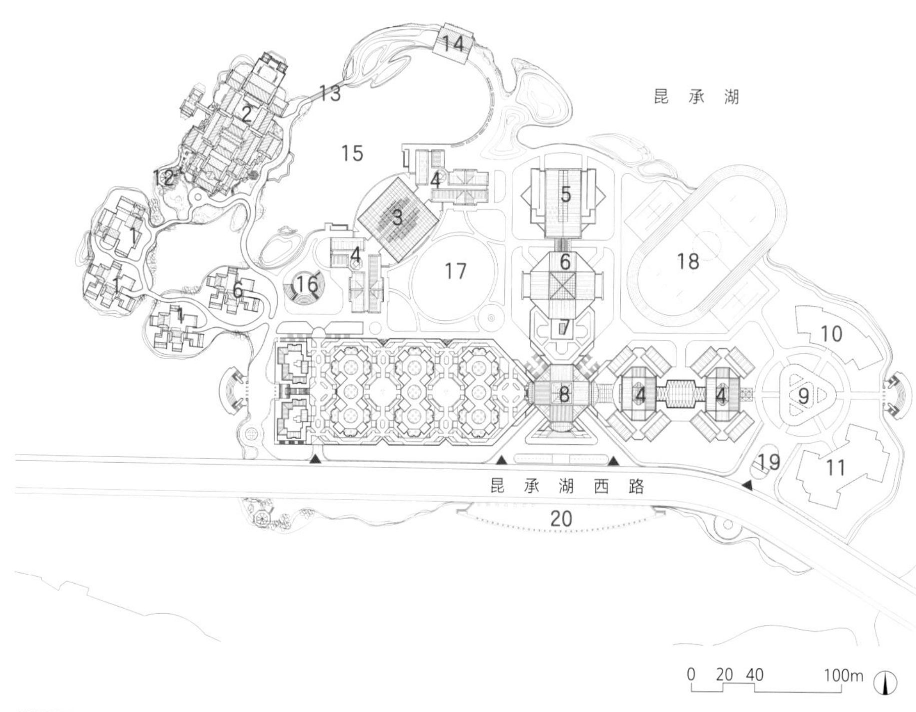

1 招待中心
2 国学书院
3 图书馆 / 餐厅
4 培训中心
5 体育中心
6 戏剧中心
7 行政中心
8 百家坛
9 二期食堂 / 图书馆
10 二期招待中心
11 二期培训中心
12 八角亭
13 驾月桥
14 和平屋 / 船屋
15 UWC 环礁湖
16 露天剧场
17 中央草坪
18 室外运动场
19 警卫室
20 国旗广场

总平面

江苏省木渎高级中学

JIANGSU PROVINCE MUDU SENIOR HIGH SCHOOL

项目类型：教育建筑　　**项目地点**：江苏苏州

用地面积：29047m²　　**建筑面积**：96270m²

设计时间：2004年　　**竣工时间**：2006年

2008年度全国优秀工程勘察设计行业奖二等奖

2009年度中国建筑学会建筑创作大奖入围奖

2006年度江苏省第十二届优秀工程设计二等奖

2006年度江苏省城乡建设系统优秀勘察设计二等奖

项目位于苏州市木渎镇著名的天平山、灵岩山风景区三级保护区内，自然风光秀美。规划设计吸收了苏州传统建筑的精华，通过围合、切割、过渡、转折、框景等空间处理手法，创造出虚实错落、富有生机和活力的空间形象。建筑材料遵循本地气候和生态的效果，主要材料有传统的砖、瓦、石，以及部分玻璃等现代材质，象征着木渎中学精致、质朴的性格特征，结合苏式建筑的黑白灰色调，营造出“烟雨江南”的意境。

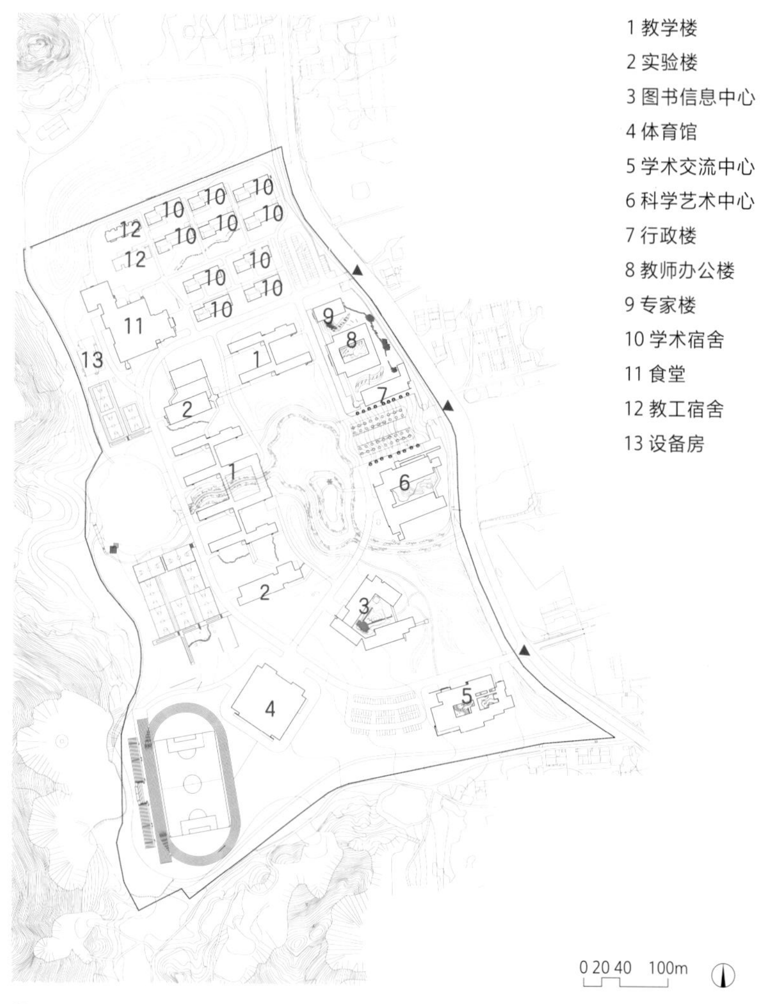

总平面

江苏省苏州实验中学科技城校

JIANGSU SUZHOU EXPERIMENTAL MIDDLE SCHOOL SCIENCE AND TECHNOLOGY CITY CAMPUS

项目类型：教育建筑　　**项目地点**：江苏苏州
用地面积：101285m²　　**建筑面积**：90362m²
设计时间：2014年　　**竣工时间**：2018年

2020年度江苏省第十九届优秀工程设计一等奖
2019年度江苏省城乡建设系统优秀勘察设计一等奖

项目位于苏州科技城，设计充分利用地块的景观条件，试图营造出自然、安全的校园环境。项目场地以自然山形肌理为主，传统阵列式排布的学校无法与周围环境形成合理的图底关系。设计师将建筑的体量顺应场地自然展开，围绕周边环境呈放射状布置，最大化地回应了场地关系。

为了实现空间的最高效利用，教学区建筑环绕山水呈聚落式展开，并采用“串葫芦”的方式，通过游廊将所有的建筑连接起来。与以往建筑围合庭院的“围城式”校园不同，在“串葫芦”式规划形成的自由多中心布局下，最大化融入山水景观资源，建筑、人、自然和谐共生。

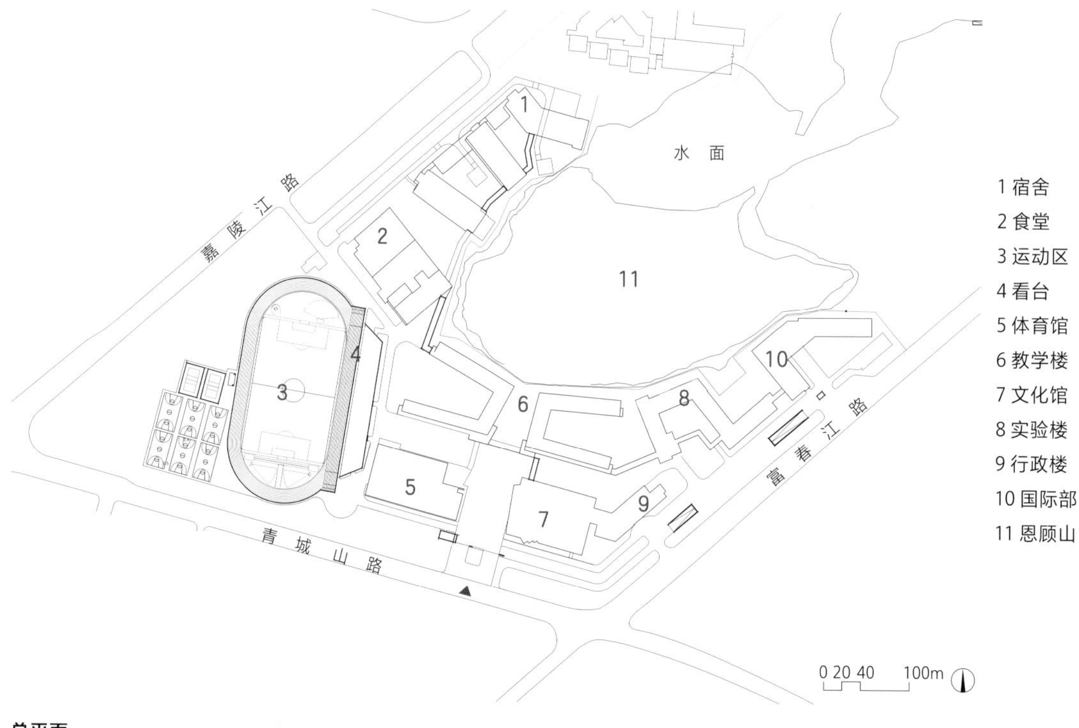

总平面

苏苑高级中学

SUYUAN SENIOR HIGH SCHOOL

项目类型：教育建筑　　**项目地点：**江苏苏州

用地面积：66908m²　　**建筑面积：**85878m²

设计时间：2019年　　**竣工时间：**2021年

2022年度江苏省第二十届优秀工程设计一等奖

2022年度江苏省城乡建设系统优秀勘察设计一等奖

项目位于苏州市吴中区，建筑高度控制为檐口高度不大于24m。学校依据江苏省普通高中四星级要求，打造吴中核心教育资源。

校园西借石湖文风，引榜眼之意，秉持传承苏式传统庭院布局精神，通过“院”“苑”“园”三大核心理念设计，将院落和建筑结合分布，同时运用现代材料演绎并结合大屋顶体现“粉墙黛瓦”的苏式意境，为在校师生创造一个富有苏式韵味的学习环境和活动环境。

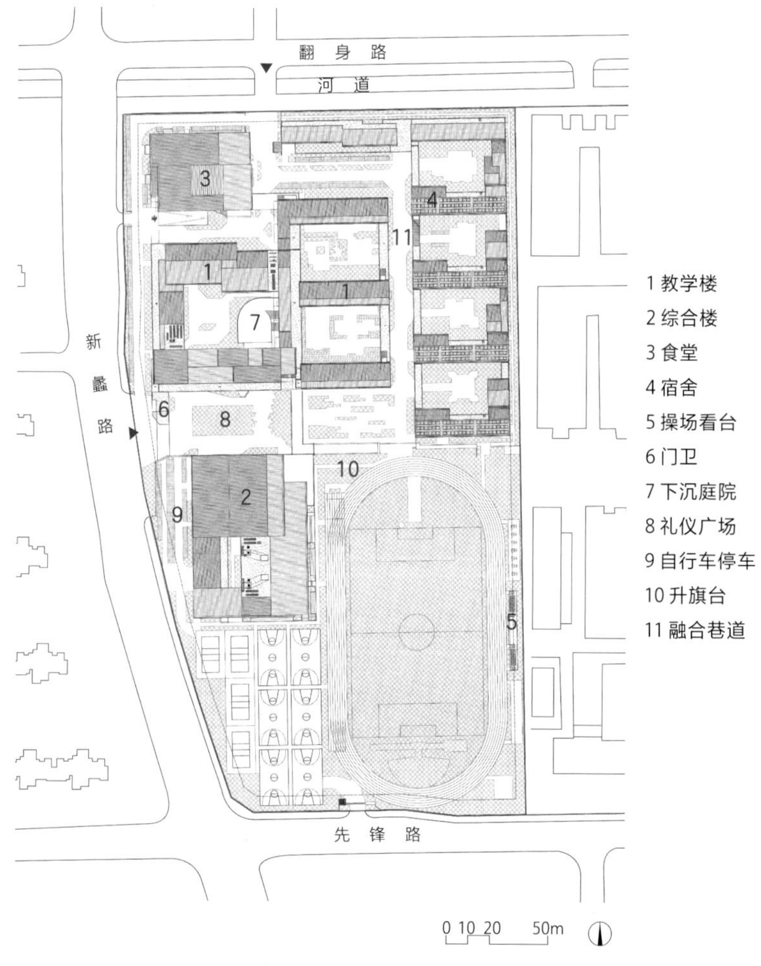

总平面

吴江中学初中部

MIDDLE SCHOOL DEPARTMENT OF WUJIANG HIGH SCHOOL

项目类型：教育建筑　　**项目地点**：江苏苏州
用地面积：41124m²　　**建筑面积**：29140m²
设计时间：2015年　　**竣工时间**：2017年

2019年度全国优秀工程勘察设计行业奖二等奖
2019年度江苏省第十八届优秀工程设计一等奖
2018年江苏省城乡建设系统优秀勘察设计一等奖
2015年度江苏省土木建筑学会第九届“建筑创作奖”三等奖

项目地处太湖之滨，拥有独特的自然地理条件和深厚的文化艺术积淀。吴江中学作为吴江区重要学府，始建于1912年，积淀了丰富的文化底蕴。

基于吴江中学的办学理念“人文立校，素质立身”，设计以人文“建”校为理念，进行了多种布局方式的尝试，权衡利弊之后甄选出最为合理的布局。同时在设计中寻找互换性的空间符号，将中国古典文化的载体——园林，作为基本的空间表达形式。设计将传统文化元素巧妙地融入学校建筑，展现学校严谨求实的治学之风，为师生营造轻松活跃的教学氛围。

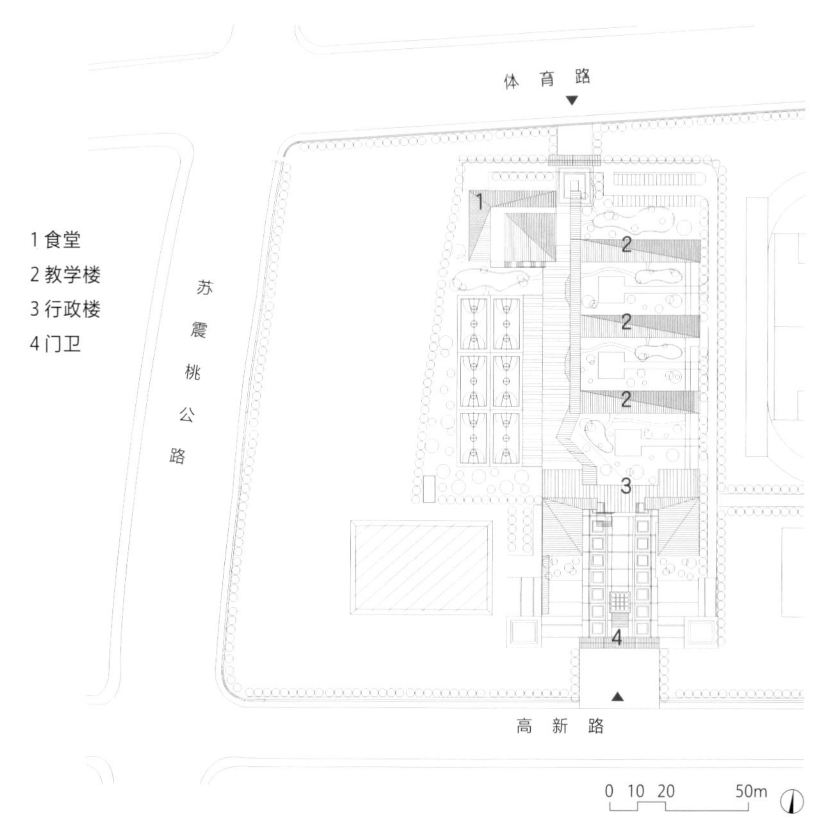

总平面

苏州高新区实验初中东校区扩建

SND EXPERIMENTAL MIDDLE SCHOOL EAST CAMPUS EXPANSION

项目类型：教育建筑　　**项目地点**：江苏苏州

用地面积：16296m²　　**建筑面积**：18326m²

设计时间：2016年　　**竣工时间**：2018年

2020年度江苏省第十九届优秀工程设计一等奖

2019年度江苏省城乡建设系统优秀勘察设计一等奖

项目位于苏州高新区狮山街道，属于在原址校园中进行扩建的类型，扩建场地极其局促，设计师在“螺蛳壳里做道场”，开启了“极限挑战”模式：在极为有限的用地中创造最大化的教学与活动空间，让师生充分感受校园综合体的魅力。

项目的功能空间体量各异、要求不同，根据特性对功能做了梳理和区分。扩建部分犹如一个“教育综合体”,有着复合的公共功能结构，提供许多共享空间作为交流场所，同时给教学提供了隐形课堂，为学校带来教育模式的革新，激发学生学习的兴趣和乐趣。建筑生成之后，出现了时空叠加的场景，场地上往日的交通功能还在，又叠加上了今日崭新的学习生活场景。各功能区的纵横交错，也形成了多维度的空间场景，超越了场地的束缚，成为了时空叠加、新旧融合的整体。

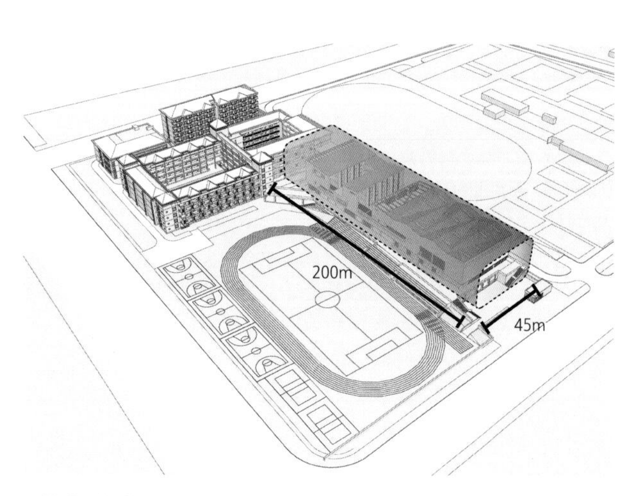

“管道”概念图

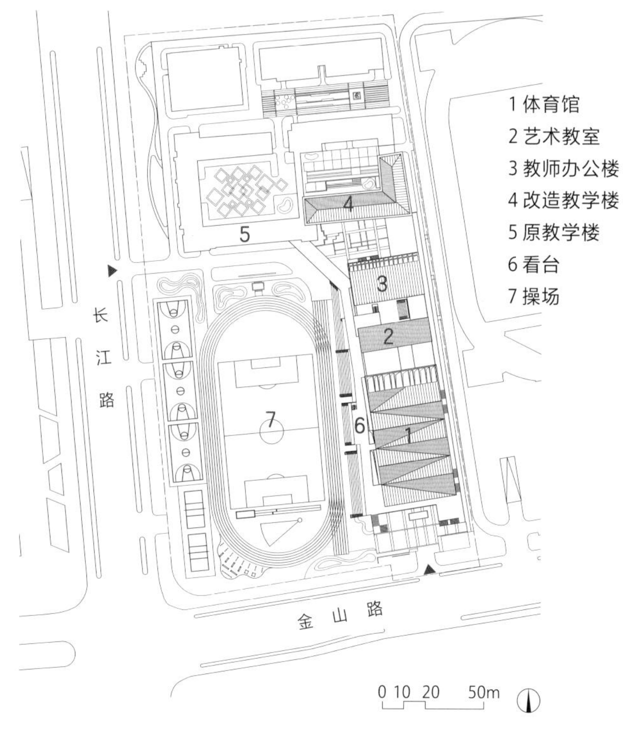

总平面

梁丰初级中学西校区

WEST CAMPUS OF LIANGFENG JUNIOR HIGH SCHOOL

项目类型：教育建筑　　**项目地点**：江苏张家港

用地面积：53555m^2　　**建筑面积**：49712m^2

设计时间：2017年　　**竣工时间**：2021年

2022—2023年度AEEDA亚洲教育环境设计奖金奖

2022年度江苏省第二十届优秀工程设计一等奖

2022年度江苏省城乡建设系统优秀勘察设计一等奖

项目位于张家港市，由包括教学楼、行政楼、报告厅、图书馆、餐厅、体育馆在内的数栋建筑组合而成。

设计借鉴传统中式书院布局，打造“一带、二轴、三中心、六合院”的空间结构。通过将局部建筑底层架空，打通校园整体景观绿化，不同尺度、风格各异的六个合院，在保证自然采光通风的同时，也给使用者提供舒适宜人的景观空间。有序、灵活的空间组合，将梁丰初级中学打造成为符合现代化教育风格的新型书院。

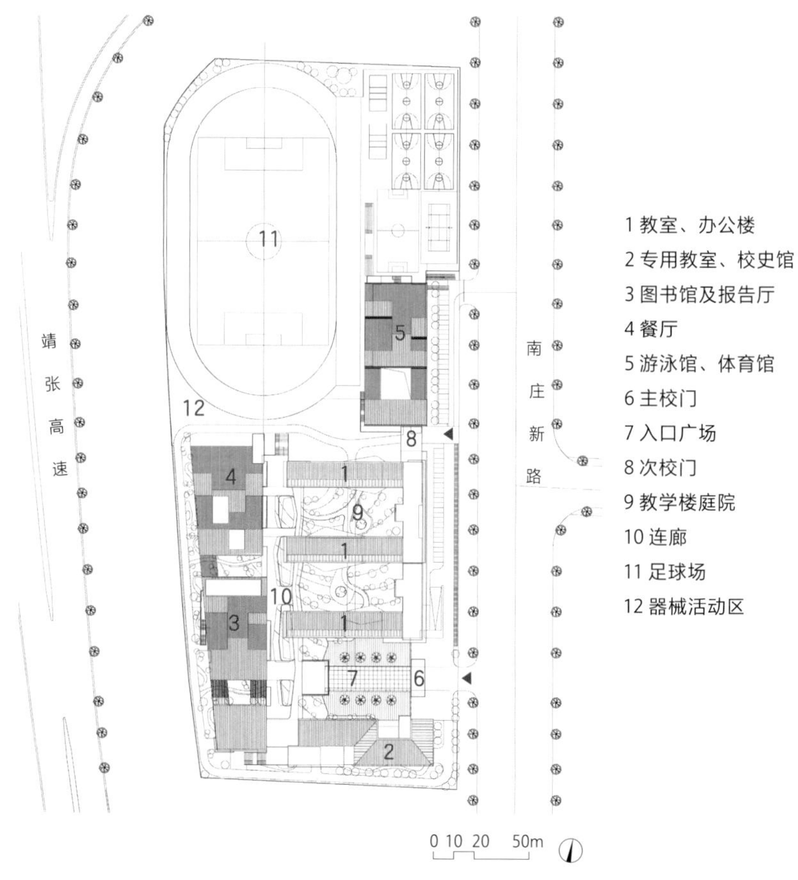

总平面

星海实验中学

XINGHAI EXPERIMENTAL MIDDLE SCHOOL

项目类型：教育建筑　　**项目地点**：江苏苏州
用地面积：66996m²　　**建筑面积**：88560m²
设计时间：2019年　　**竣工时间**：2021年

2022年度江苏省第二十届优秀工程设计三等奖
2022年度江苏省城乡建设系统优秀勘察设计三等奖

项目位于苏州工业园区核心区域，周边环绕林立着商业大楼和高层住宅。对于快速发展的城市中心地带，每寸土地都打着经济效率的标签；但学生需要的却是一片能够自由奔跑、多元化发展的空间。

设计面临的挑战是如何解决有限的用地资源和学校对多样化空间需求之间的矛盾。为了使建筑空间更加人性化，场地和空间的设计尽可能主动地贴合教学与活动的需求，在课堂内外带给学生一处轻松学习生活的场所。集合化的功能布局、多层次的活动空间、立体式的交通网络、沉浸式的景观庭院成为本项目的设计亮点。

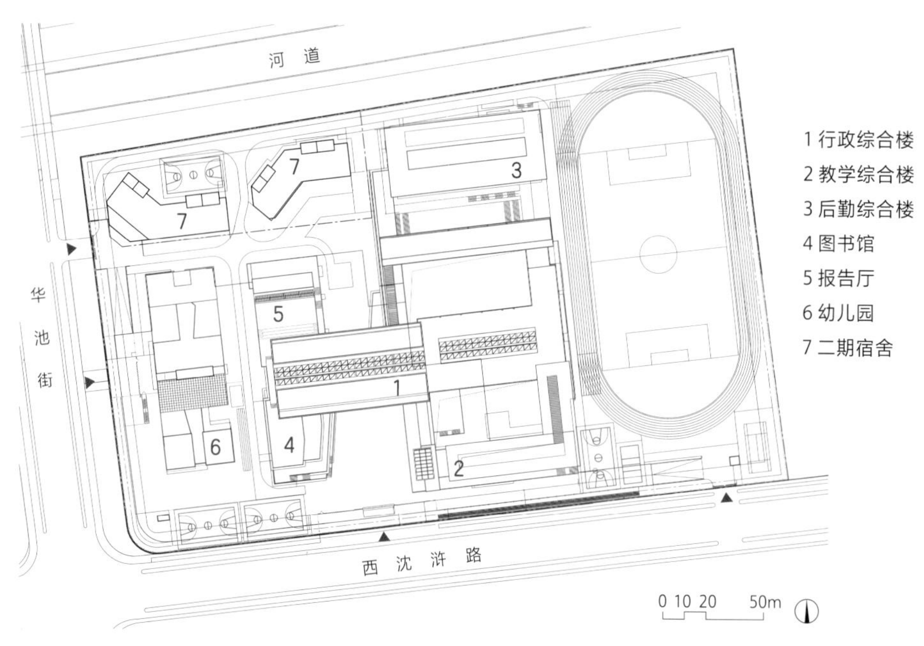

总平面

星耀广场

常熟市三环小学

CHANGSHU SANHUAN PRIMARY SCHOOL

项目类型：教育建筑　　**项目地点**：江苏常熟
用地面积：46246m²　　**建筑面积**：48293m²
设计时间：2016年　　**竣工时间**：2020年

2021年度江苏省城乡建设系统优秀勘察设计二等奖
2022年度江苏省第二十届优秀工程设计三等奖
2019年度江苏省土木建筑学会第十三届“建筑创作奖”二等奖

项目位于江苏省常熟市南部新城，坐落于昆承湖之滨，设计注重与城市的空间关系的协调。为了避免互相干扰，小学部分和幼儿园部分被相对独立地布置。

设计着重强调绿色低碳的理念，致力于创造一个环保的校园，采取景观绿化和花园式校园设计等多种措施，通过植入自然元素和相互渗透的景观布局，创造出宜人的学习环境，并倡导节能减排的理念。

设计在保留苏式坡屋顶的基础上进行了创新，将主入口处的建筑退让和抬高，形成了一个宽敞的入口广场空间。屋顶的设计灵感来源于翻起的书页和太湖中的扁舟，通过白墙、混凝土和红砖的搭配，并点缀丰富的色彩，创造了一个充满活力且富有文化底蕴的学校建筑形象。

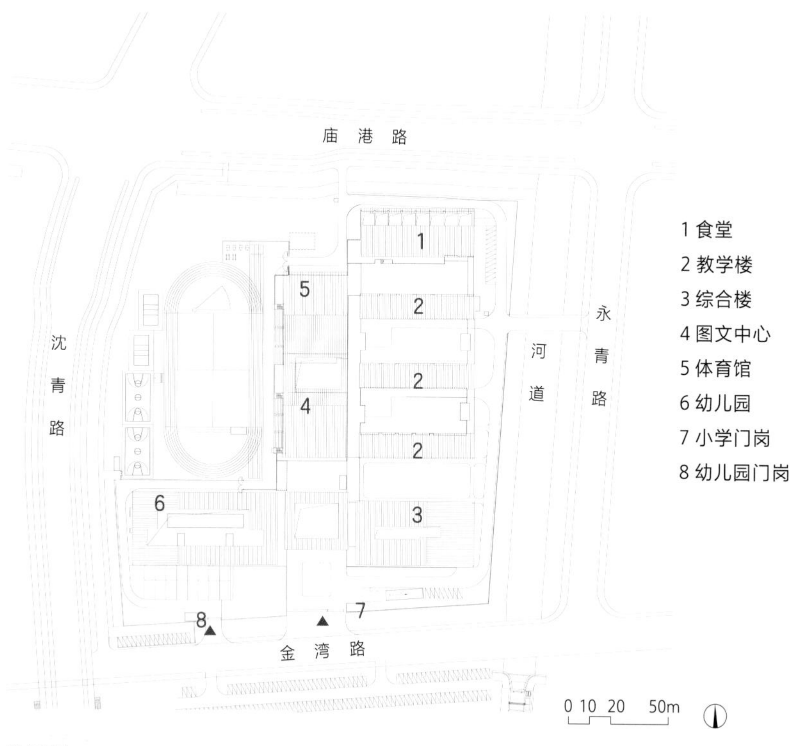

1 食堂
2 教学楼
3 综合楼
4 图文中心
5 体育馆
6 幼儿园
7 小学门岗
8 幼儿园门岗

总平面

吴郡幼儿园

WUJUN KINDERGARTEN

项目类型：教育建筑　　**项目地点**：江苏苏州

用地面积：12527m²　　**建筑面积**：20266m²

设计时间：2017年　　**竣工时间**：2018年

2020 ASIA DESIGN PRIZE winner

2020年度江苏省第十九届优秀工程设计三等奖

2020年度江苏省城乡建设系统优秀勘察设计二等奖

2019年度江苏省土木建筑学会第十三届建筑创作奖一等奖

项目是苏州太湖新城吴郡片区的第一座幼儿园，东南方向毗邻东太湖水面，环境舒适，景色优美。片区里所有的居住建筑色调偏灰白和土黄，设计为这个湖边的环境加入一抹亮色，给这个社区增加更多的生气，让居民从周边高楼上看下来，能立刻找到这个以幼儿园为"注点"的滨湖空间，而不至于迷失在灰白和土黄的钢筋水泥的"森林"里。

项目用地轮廓是一个直角梯形的形状。采用三排三层的建筑体量，用一条南北贯穿的连廊将三排教室连接起来，并将三排教室和连廊整体在平面上旋转约10°，连廊基本与东边界平行，解决了连廊的自遮挡问题，扩大了最北侧教室楼与入口前道路的距离，改善了幼儿园入口的空间尺寸。

幼儿园的外立面采用铝板外饰面。通过对于铝板和土建墙身进退尺寸的精确控制，建筑外立面上实现了丰富的层次变化和细节效果。

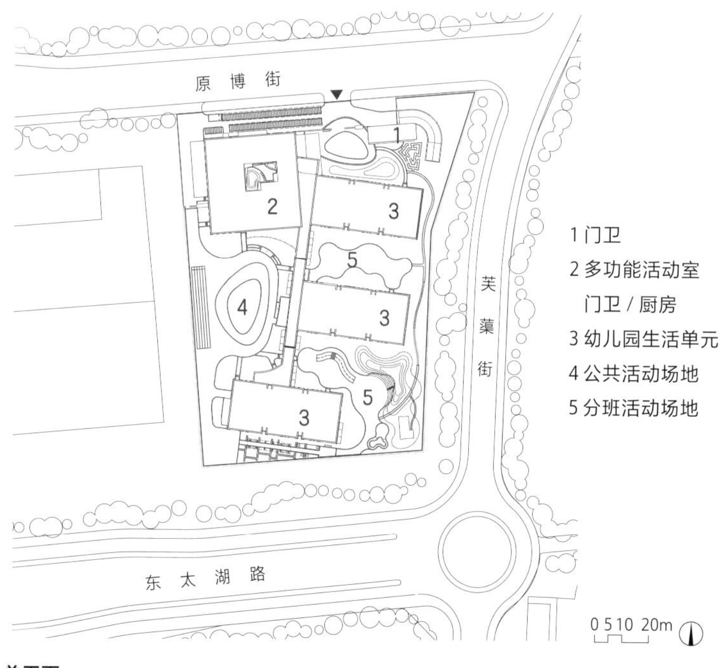

1 门卫
2 多功能活动室
门卫 / 厨房
3 幼儿园生活单元
4 公共活动场地
5 分班活动场地

总平面

重庆哈罗国际学校

HARROW CHONGQING

项目类型：教育建筑　　**项目地点**：重庆
用地面积：95616m²　　**建筑面积**：61734m²
设计时间：2018年　　**竣工时间**：2020年

2022—2023年度AEEDA亚洲教育环境设计奖银奖
2022年度江苏省第二十届优秀工程设计三等奖
2022年度江苏省城乡建设系统优秀勘察设计三等奖

项目位于重庆两江新区的英伦风情小镇西北角，基地西南高、东北低，自然坡道0～10°，呈现出重庆地区的典型地貌特征。地块的高差给场地设计与建筑布局带来了挑战，重庆充满高差变化的场地特征将造成师生日常上下课的爬坡问题，而如何利用和化解高差，缩短师生在场景切换中的必要流线，成为场地设计的关键。设计利用地势，营造了通往不同功能区的多标高入口，对场地高差条件和土方量进行梳理后，结合学校各功能空间特性与规范，将场地划分为3阶标高，形成3个片区，一方面保证了场地土方量的经济性，另一方面将场地高差集中于主要轴线处。

在建筑形式上，结合其学校自身的历史渊源，延续欧式建筑特色。在标志性坡屋顶立面的基础上进行简化，并通过平坡结合的手法处理大体量空间的屋顶形式，化整为零地消解了屋顶庞大的体量感，使整个校园既具有英伦校园的古典氛围，又具有现代教育建筑的简约特色。

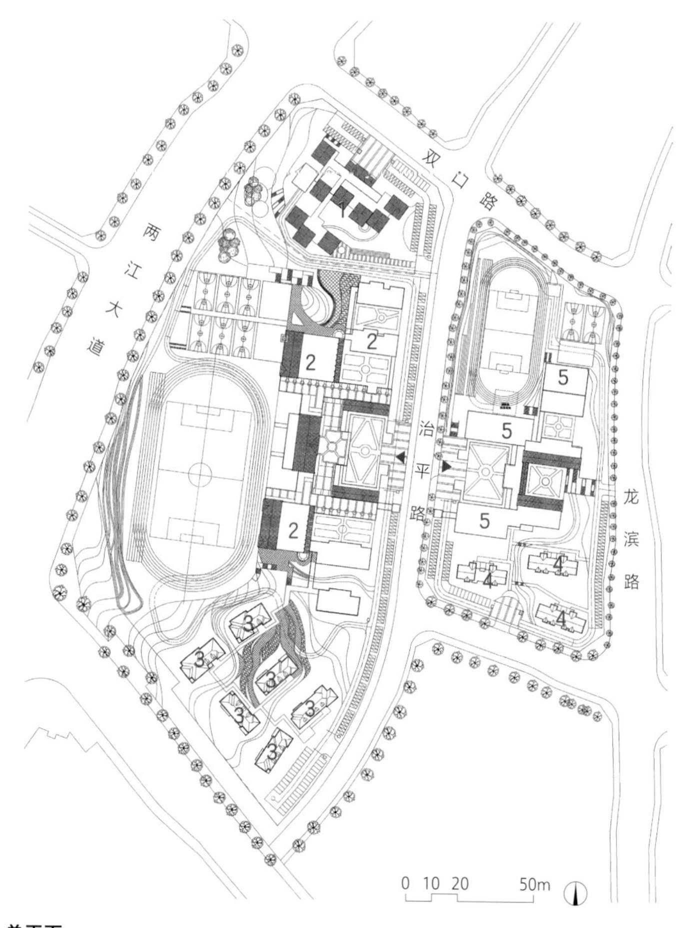

总平面

HARROW CHONGQING

伊顿国际学校

ETON HOUSE INTERNATIONAL SCHOOL

项目类型：教育建筑　　**项目地点**：江苏苏州

用地面积：29047m²　　**建筑面积**：12365m²

设计时间：2007年　　**竣工时间**：2008年

2009年度全国优秀工程勘察设计行业建筑工程设计二等奖

2010年度江苏省第十四届优秀工程设计二等奖

2009年江苏省城乡建设系统优秀勘察设计二等奖

基于国际学校个性化、多元化的教育诉求，采用围合式的空间营造，用连廊连接各个相对独立的功能区，形成外部开放空间和内部扩展空间，使教学功能与外部环境有机地联系起来，更有利于公共活动的开展，鼓励聚集，促进交流，在校园环境中塑造出充满活力、内外渗透的复合型教育空间。

建筑体量保持一个较为亲切的尺度，造型兼顾现代与传统，立面构图处理体现学校活跃的氛围，外墙以红砖为主，局部点缀白墙，通过细节处理刻画建筑体块，形成了东西方文化的对话。

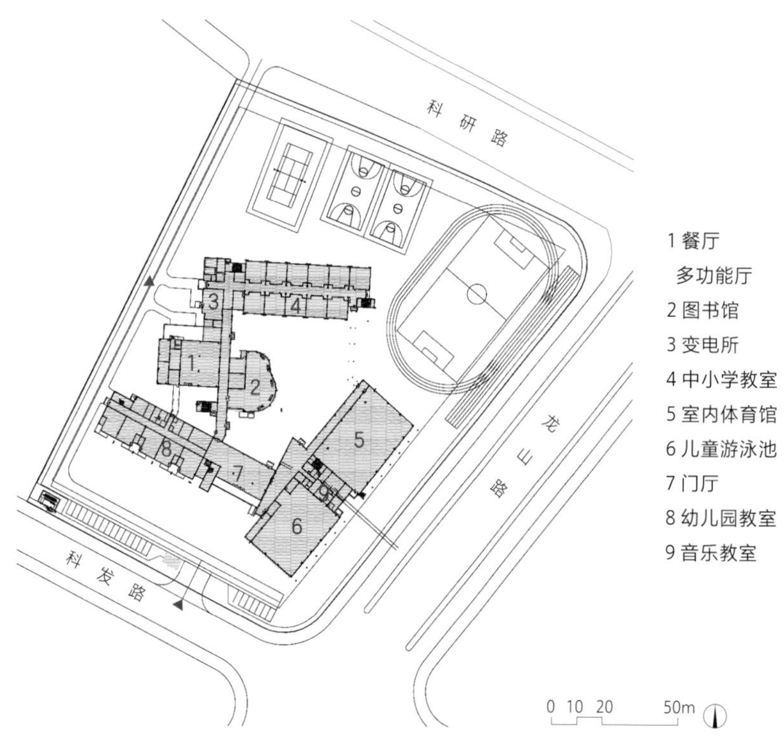

总平面

苏州伊顿国际学校
EtonHouse

07

医养建筑

MEDICAL AND ELDERLY CARE ARCHITECTURE

响水县人民医院新院区、妇幼保健院
XIANGSHUI PEOPLE'S HOSPITAL NEW CAMPUS MATERNAL AND CHILD HEALTH HOSPITAL
苏州大学附属第二医院应急急救与危重症救治中心
EMERGENCY AND CRITICAL CARE CENTER OF THE SECOND AFFILIATED HOSPITAL OF SOOCHOW UNIVERSITY
张家港中医院
ZHANGJIAGANG HOSPITAL OF TRADITIONAL CHINESE MEDICINE
苏州吴中区公共卫生中心
PUBLIC HEALTH CENTER OF WUZHONG DISTRICT
昆山公共卫生中心
KUNSHAN PUBLIC HEALTH CENTER
苏州市立医院康复医疗中心
REHABILITATION MEDICAL CENTER OF SUZHOU MUNICIPAL HOSPITAL
国寿雅境二期项目
GUOSHOU YAJING PHASE Ⅱ PROJECT
苏州工业园区久龄公寓
SIP JIULING APARTMENT
苏州市怡养老年公寓
SUZHOU YIYANG SENIOR APARTMENT
苏州西园养老护理院
SUZHOU XIYUAN PENSION AND NURSING HOME

响水县人民医院新院区、妇幼保健院

XIANGSHUI PEOPLE'S HOSPITAL NEW CAMPUS MATERNAL AND CHILD HEALTH HOSPITAL

项目类型： 医疗建筑　　**项目地点：** 江苏盐城

用地面积： 210488m^2　　**建筑面积：** 281000m^2

设计时间： 2019年　　**竣工时间：** 在建

床位数量： 1280　　**合作单位：** 山东省建筑设计研究院有限公司

项目位于响水县，黄海之滨，灌河南岸，由门诊楼、病房楼、妇幼保健院、传染病楼、行政楼等数栋医疗建筑组合而成，是当地唯一的一座三级综合性医院。

设计以人为本，从建筑布局、环境设施到诊治服务的全过程均以方便患者为核心，给人以便捷、舒适的就医环境。建筑采用通透的立面，塑造出简洁灵动的设计风格。通过丰富的细节表现现代化医院实用、舒适的特征。同时在建筑设计中营造绿色生态的景观环境，采用高效的能源利用设施，结合可再生能源，减少环境污染并节约能源。

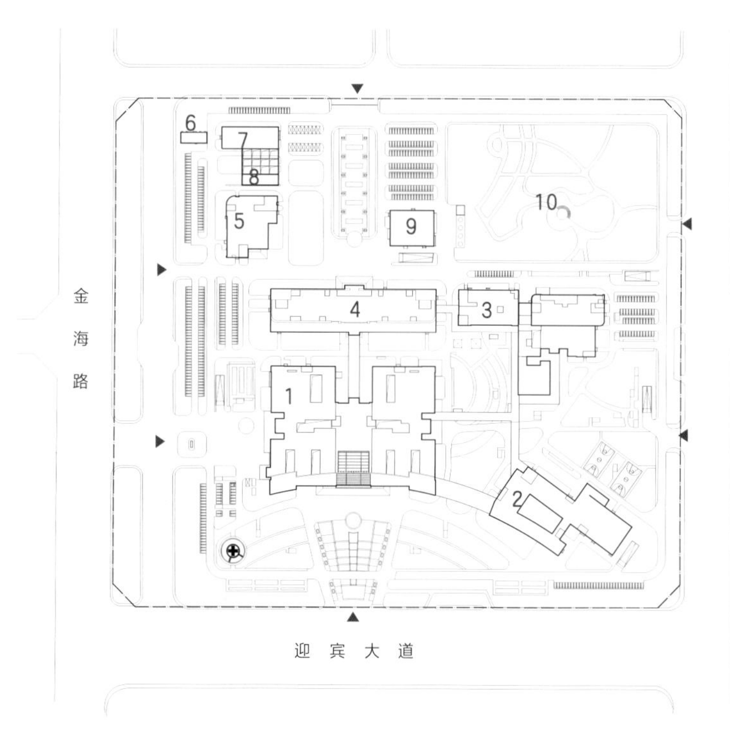

1 门诊医技楼
2 行政科教综合楼
3 妇幼保健院
4 病房楼
5 传染病楼
6 污水处理站
7 120 急救中心
8 洗消中心
9 洗衣房、垃圾暂存间
10 预留发展用地（健身设施场地）

0 20 40 100m

总平面

妇幼保健院

苏州大学附属第二医院应急急救与危重症救治中心

EMERGENCY AND CRITICAL CARE CENTER OF THE SECOND AFFILIATED HOSPITAL OF SOOCHOW UNIVERSITY

项目类型：医疗建筑　　**项目地点**：江苏苏州

用地面积：8000m^2　　**建筑面积**：92006m^2

设计时间：2020年　　**竣工时间**：在建

本项目定位是建设为国际标准、国内领先的智慧型军民两用应急急救与危重症救治国家临床中心。项目与相邻地块的苏州市急救中心项目统一规划设计和整体建设。设计难点在于有限场地环境下系统合理地解决复杂的医疗功能与流线组织，在此基础上着力构建符合地域文脉的现代建筑空间和形式。建筑采用集中式布局，充分利用地下、裙房及主楼空间特点合理布置功能分区，科学组织各类功能流线，增加绿化内庭院提高建筑的自然通风和采光，巧妙利用因日照影响产生的层层退台构建立体园林景观，同时提升患者就诊效率，也为医护人员提供便捷舒适的工作环境。

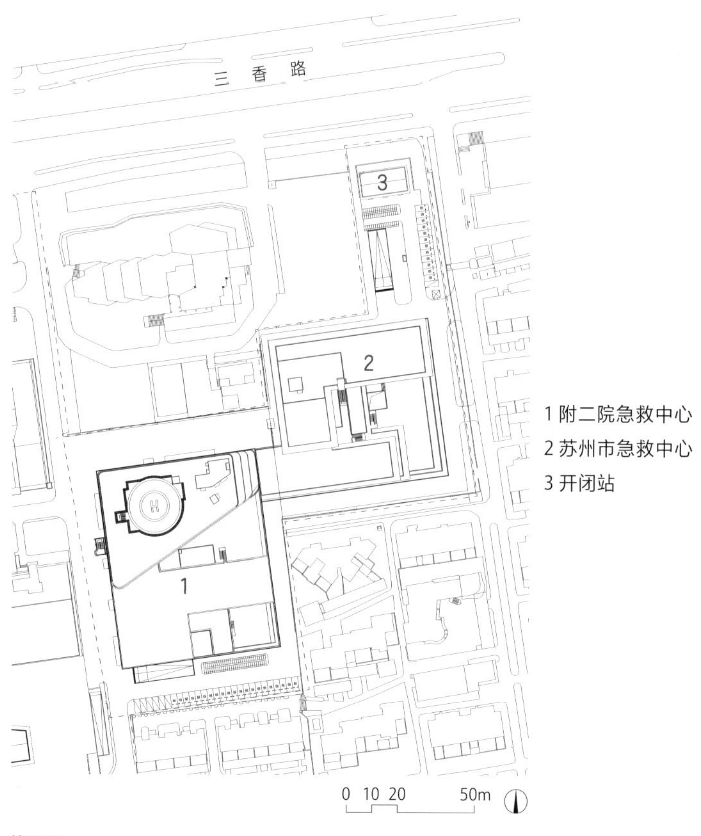

1 附二院急救中心
2 苏州市急救中心
3 开闭站

总平面

苏州市急救中心

张家港中医院

ZHANGJIAGANG HOSPITAL OF TRADITIONAL CHINESE MEDICINE

项目类型：医疗建筑　　**项目地点**：江苏张家港
用地面积：105797m²　　**建筑面积**：185000m²
设计时间：2022年　　**竣工时间**：在建
床位数量：1500

项目定位为集医疗、教学、科研、保健、康复为一体的三级甲等综合性中医医院，满足6000人次日门急诊量需求。设计以“园林中的医院”为构思，建筑造型从江南传统建筑特色中汲取灵感，整体色彩选用灰白色作为主色调，通过连续起伏的坡屋面，采用现代设计手法对传统建筑语汇进行重塑，营造出轻盈而富有变化的中式立面效果。住院楼立面采用光伏遮阳一体化的设计手法，以防止眩光，减少能耗；地面采用透水铺装，以减少雨水地面径流；建筑塔楼屋面采用光伏分布式能源系统及太阳能热水系统，加强对绿色能源的利用。

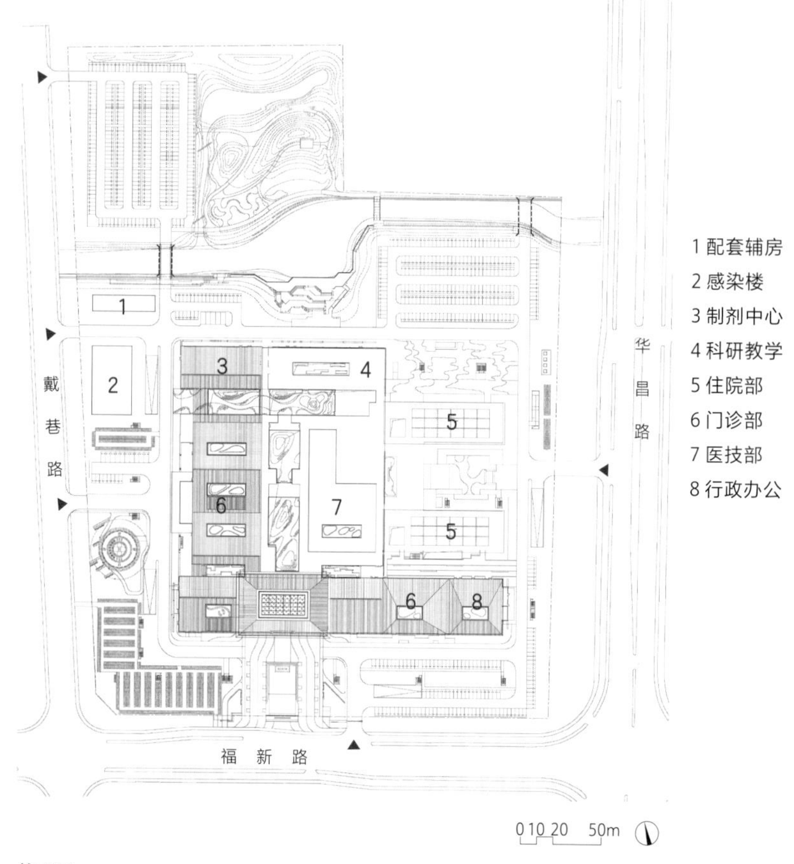

总平面

苏州吴中区公共卫生中心

PUBLIC HEALTH CENTER OF WUZHONG DISTRICT

项目类型：医疗建筑　　**项目地点**：江苏苏州
用地面积：15748m^2　　**建筑面积**：34510m^2
设计时间：2022年　　**竣工时间**：在建

项目满足区域内疾病防控、卫生监督、妇幼保健、应急指挥、卫生信息发布等公共卫生服务的需要，为广大群众的健康安全保障提供更加坚实的基础和技术支撑。

项目遵循“以人为本、集约高效、智能现代、绿色生态”的设计理念。设计充分体现建筑功能需求，着重表现建筑风格及地域、文化特点，满足区域公共卫生服务需求，打造具有人性化、智能化的现代公共卫生中心。

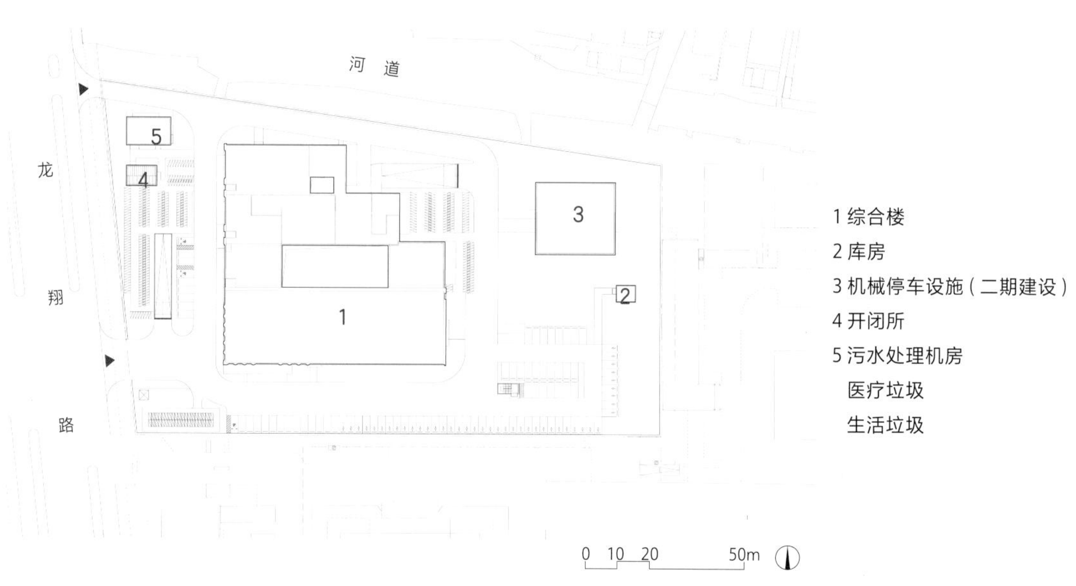

总平面

昆山公共卫生中心

KUNSHAN PUBLIC HEALTH CENTER

项目类型：医疗建筑　**项目地点**：江苏苏州
用地面积：20117m^2　**建筑面积**：72428m^2
设计时间：2016年　**竣工时间**：2021年

2022年度江苏省城乡建设系统优秀勘察设计二等奖
2022年度江苏省第二十届优秀工程设计三等奖

项目作为昆山重点实事工程“三大医疗中心”之一，是昆山公共卫生事业发展的重要里程碑。项目以公共卫生健康服务为切入点，从人民群众根本需求出发，探索一个能够为市民提供健康教育、疾病预防控制、无偿献血、公共卫生应急管理等一站式公共卫生服务的综合解决方案。

建筑与城市之间的互动是设计需要考虑的重点。通过两栋塔楼加裙房的布置，使各个职能部门既独立又关联，方便市民办理业务，缩短流线距离。设计结合昆山境内水网密布、河路并行的特点，充分挖掘昆山当地文化内涵，以“流动的水”作为形体设计意象。流线型的裙房蜿蜒连绵，风帆造型的主楼生机盎然，寓意生命延续、生生不息。

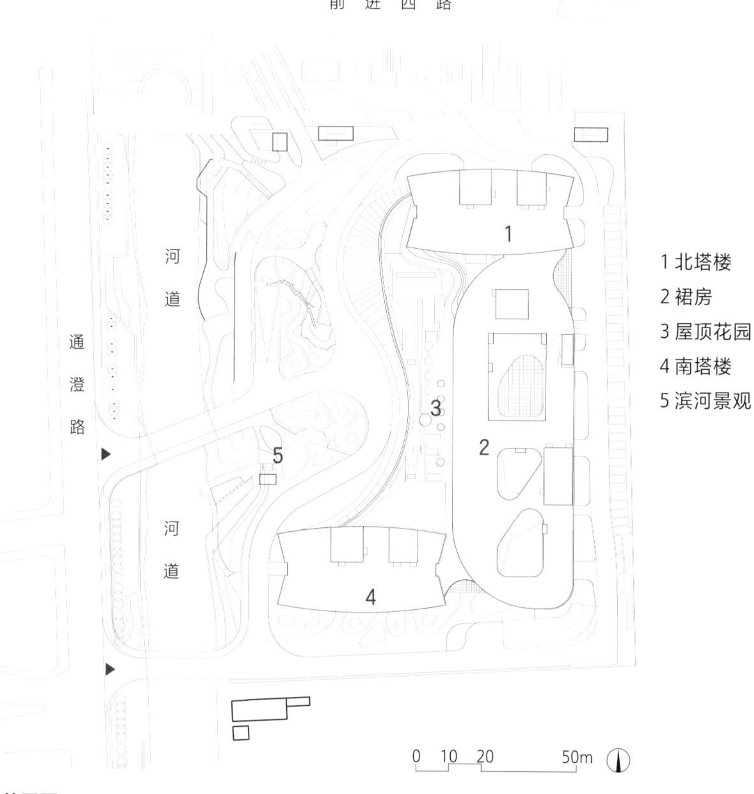

总平面

苏州市立医院康复医疗中心

REHABILITATION MEDICAL CENTER OF SUZHOU MUNICIPAL HOSPITAL

项目类型：医疗建筑　　**项目地点**：江苏苏州

用地面积：25637m^2　　**建筑面积**：69881m^2

设计时间：2020年　　**竣工时间**：在建

床位数量：400

项目位于苏州市姑苏区，在姑苏城外护城河边，为广济医院旧址，是一块狭长不规则洼地。北侧毗邻苏州火车站，东南面是滨河绿化带及护城河，风景优美。

因项目用地位于古城区特殊节点，如何在严苛的设计条件下，布局一个既尊重城市肌理、又满足院方建设规模及康复疗愈需求的建筑，同时处理好医疗建筑复杂的流线关系，是设计的最大难点和重点。

建筑整体上呈两个高低体量并行布置，满足规划限制高度，布局顺应地形伸展开来，使病房更易获得最佳日照，拥有最开阔的护城河景观视野，同时形成中轴对称的建筑形态。建筑从苏州传统建筑中提取灵感，并结合现代的处理方式将其优化，苏式韵味藏于建筑的意形之间，与古城风貌和谐相融。

总平面

苏州市立医院
康复医疗中心

国寿雅境二期项目

GUOSHOU YAJING PHASE Ⅱ PROJECT

项目类型：养老建筑
用地面积：171962m²
设计时间：2020年
床位数量：437户自理型
648床自理型
259床护理型

项目地点：江苏苏州
建筑面积：142835m²
竣工时间：在建
合作单位：中国建筑设计研究院有限公司

项目位于阳澄湖半岛，一二期整体构成了阳澄湖度假区的养老养生功能。二期项目分为东西两个地块，其中东地块主要功能为养老楼和配套服务楼——国寿书院。西地块主体功能为“客养”，包括养生酒店、健康管理中心、康体会议中心等。

二期规划与一期协同，交通上以大小环线、功能互补的方式形成配套环。在空间上分为九大组团，疏密有致。建筑内外动静分区，内部庭院私密，外部活动空间开放。项目集绿色、健康、适老于一体，为全龄老人提供全生命周期的服务。

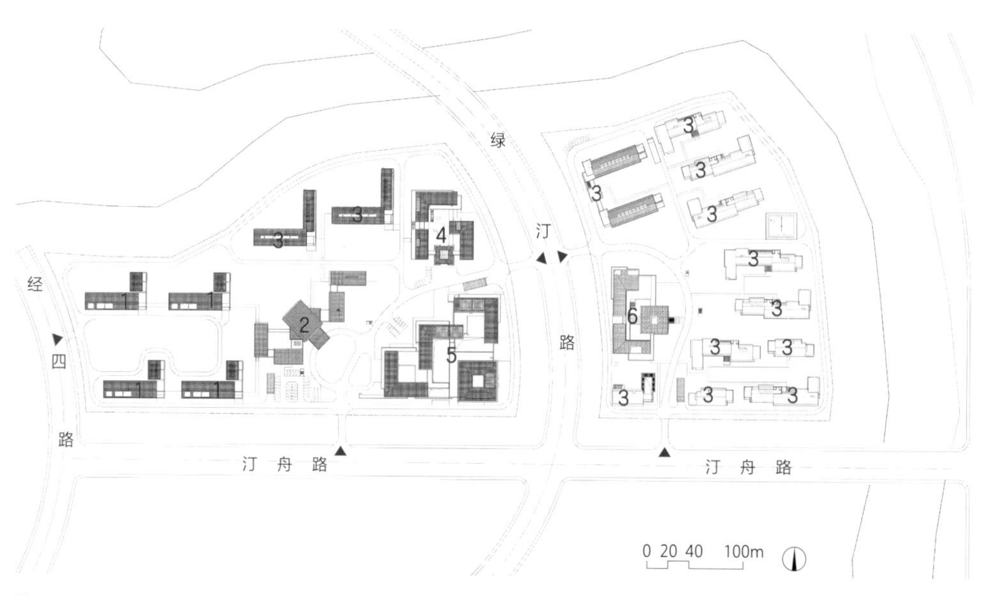

总平面

苏州工业园区久龄公寓

SIP JIULING APARTMENT

项目类型：养老建筑　　**项目地点**：江苏苏州
用地面积：32098m^2　　**建筑面积**：60158m^2
设计时间：2013年　　**竣工时间**：2016年
床位数量：养老公寓400套
护理床位250床

2017年度江苏省城乡建设系统优秀勘察设计三等奖

项目是苏州工业园区首个公立综合型养老社区，是根据园区养老服务需求而建设的医养融合服务载体。项目西北面是阳澄湖畔的高尔夫球场，东面是青剑湖，南面是星湖医院，风景优美、配套优良。

项目包括综合楼、配套服务楼、餐厅、普通养老单元、养老公寓等几大功能区块，在规划上与周边道路、建筑形成统一的形态肌理，又与城市天际线及区域规划相协调。场地设计以“怡然自得”为愿景，通过精心设计营造的优美环境，让老人感受老有所养、老有所学、老有所为、老有所乐的关怀。设计重点关注老人行为模式，在每个细节的设计都考虑使用的便利性，以打造多层次的公共空间和开放空间为主，鼓励老人走向室外，在丰富的交往空间和活动配套场所拥抱自然，享受健康人生。

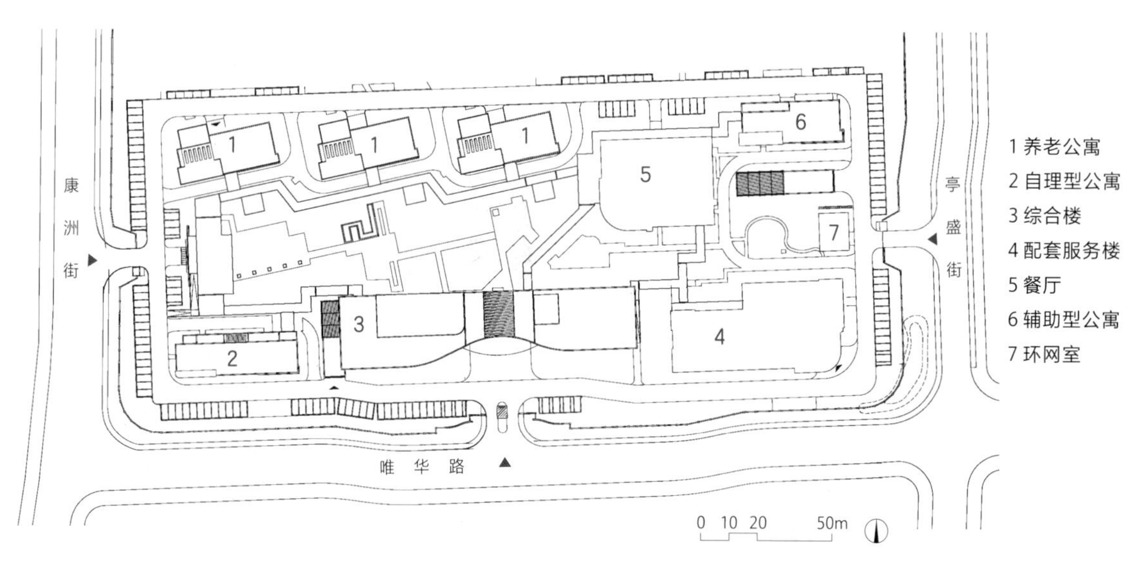

总平面

苏州市怡养老年公寓

SUZHOU YIYANG SENIOR APARTMENT

项目类型：养老建筑　　**项目地点**：江苏苏州
用地面积：44314m²　　**建筑面积**：81592m²
设计时间：2012年　　**竣工时间**：2014年

2015年度江苏省第十七届优秀工程设计三等奖
2015年度江苏省城乡建设系统优秀勘察设计三等奖

项目于苏州高新区狮山板块，是苏州第一座CCRC模式的养老社区。主体建筑共7栋，其中4栋自理型老年公寓楼、2栋护理楼、1栋配套服务楼。每栋楼由连廊连接。设计从前期策划入手，建筑功能满足全龄老人的需求，从自理、护理到临终环节，体现服务的重要性和人性关怀。

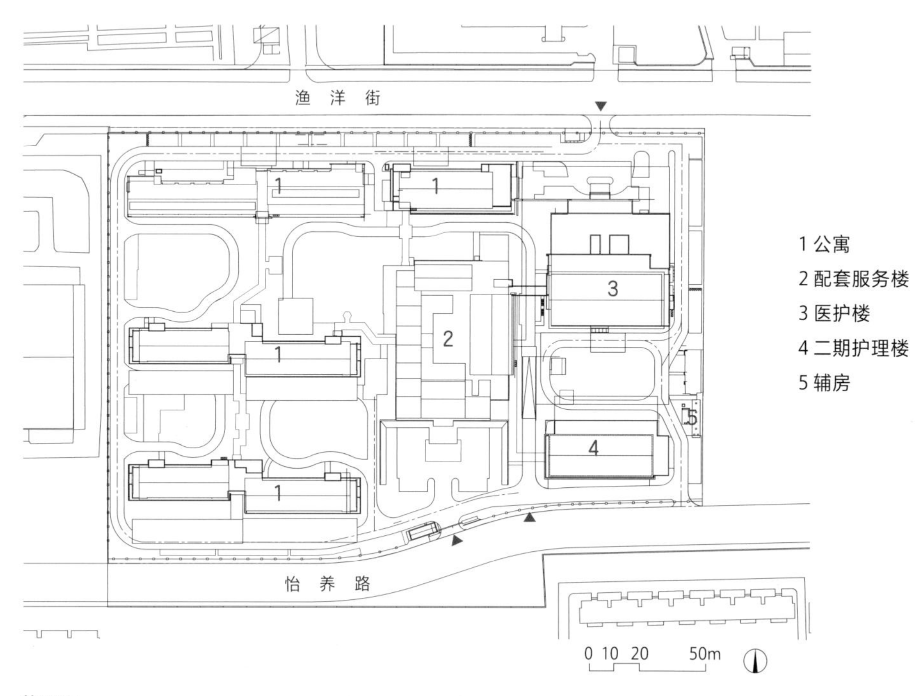

总平面

苏州西园养老护理院

SUZHOU XIYUAN PENSION AND NURSING HOME

项目类型：养老建筑　　**项目地点**：江苏苏州
用地面积：19947m²　　**建筑面积**：28256m²
设计时间：2015年　　**竣工时间**：2020年

2020年度江苏省土木建筑学会第十四届“建筑创作奖”一等奖
2021年度江苏省城乡建设系统优秀勘察设计三等奖

项目位于苏州古城区，毗邻千年佛教古刹——西园戒幢律寺。设计的总体布局在结合传统建筑的生长形态以及建筑与庭院空间比例的基础上进行城市肌理的织补，形成与西园寺融为一体的城市图底。

设计源于我国古代禅养理念，采用传统与当代美学相结合的手法创造建筑、园林、景观，旨在表达东方隐逸思想的智慧与精神。项目在总体布置上呼应古刹布局，形成内外禅院结构。对外以房为墙，在古城中围合出一个相对静谧祥和的精神空间；对内仿照院落形式，利用各个组团形成各自的庭院空间，保证每个养老单元的相对独立，实现了“隐于古城，融于古寺”的空间意境。

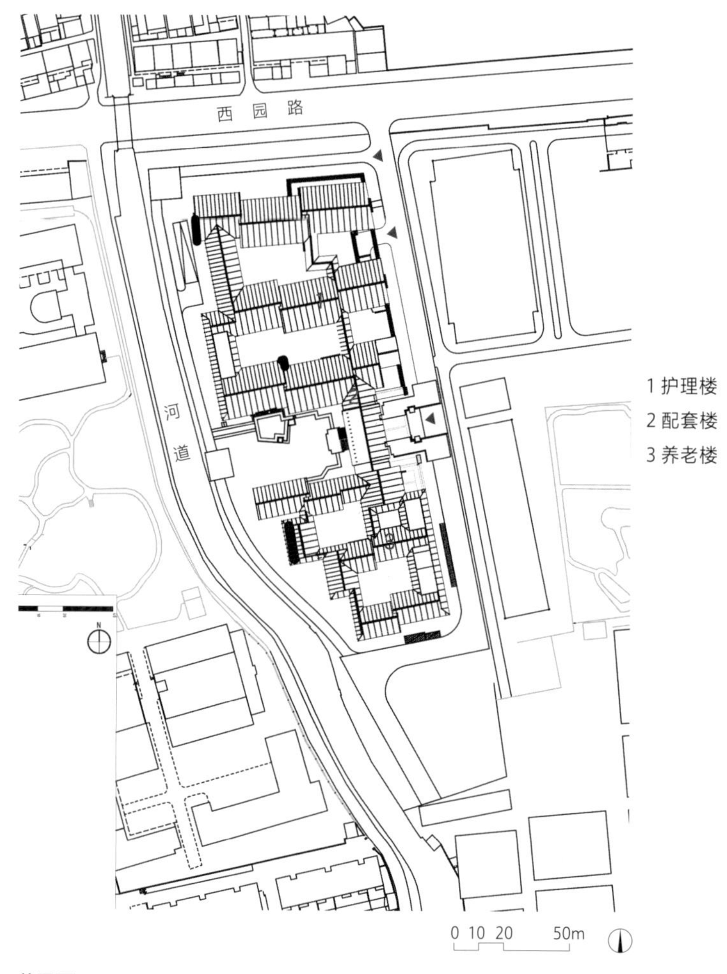

总平面

08

科技研发

SCIENTIFIC R&D ARCHITECTURE

微软（中国）苏州科技园区二期
MICROSOFT(CHINA) SUZHOU SCIENCE PARK PHASE Ⅱ
枫桥工业园
FENGQIAO INDUSTRIAL PARK
金唯智基因组研究和基因技术应用实验楼
GENEWIZ GENOME RESEARCH AND GENE TECHNOLOGY APPLICAITION LABORATORY BUILDING
骊住科技新工厂项目
LIXIL TECHNOLOGY NEW FACTORY
大兆瓦风机新园区项目
ENVISION GROUP NEW PARK PROJECT
苏州纳米城
NANOPOLIS SUZHOU
宝时得中国总部一期
POSITEC CHINA HEADQUARTERS PHASE Ⅰ
伯乐蒂森焊接技术苏州厂房
BOHLER WELDING GROUP WORKSHOP IN SUZHOU
智创山谷总部园区
ZHICHUANG VALLEY HEADQUARTERS PARK
嘉兴微创园
JIAXING MICROPORT PARK

微软（中国）苏州科技园区二期

MICROSOFT(CHINA) SUZHOU SCIENCE PARK PHASE Ⅱ

项目类型：科技园区　　**项目地点：**江苏苏州

用地面积：41446m^2　　**建筑面积：**34000m^2

设计时间：2020年　　**竣工时间：**在建

项目位于苏州工业园区国际科技园五期微软总部现址，北侧是苏州大学，南侧为纳米生物科技园。设计旨在打造一个高品质、开放和创意的办公大楼，完善苏州微软办公总部功能配置，满足目前需求的同时也能够适应未来的变化。

园区整体布局理念强调与科技园五期已有规划、建筑的对话。在充分研究周边整体空间格局的基础上，运用轴线呼应和景观渗透，与区域规划、设计肌理有机融合，采用将场地绿化延续到建筑首层屋顶的设计理念，以层层退台花坛的手法，打造层次丰富的园林景观，使建筑置身自然生态的环境之中。

项目秉承绿色建筑设计理念，从安全耐久、健康舒适、生活便利、资源节约、环境宜居等方面进行设计。着重提升建筑的绿色性能，并通过对可再生能源的合理利用，强化建筑的节能和减碳效果，最终实现LEED金级、二星级绿色建筑建设目标。

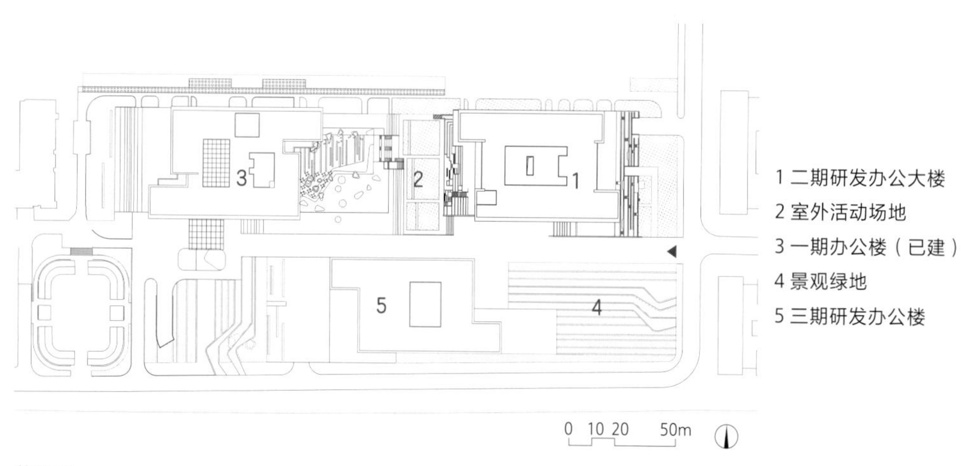

总平面

Microsoft

枫桥工业园

FENGQIAO INDUSTRIAL PARK

项目类型：科技园区　**项目地点**：江苏苏州
用地面积：51940m²　**建筑面积**：141085m²
设计时间：2017年　**竣工时间**：2021年

2022年度金拱奖——产业园区规划设计金奖
2022年度江苏省第二十届优秀工程设计一等奖（一期）
2022年度江苏省第二十届优秀工程设计一等奖（二期）
2021年度江苏省城乡建设系统优秀勘察设计一等奖（一期）
2022年度江苏省城乡建设系统优秀勘察设计一等奖（二期）

项目是苏州高新区“退二优二”发展战略的标杆示范项目。致力于打造长三角生物与生命科技产业高地，成为产业升级和城市更新形象展示区。

整个园区根据供地时序，采用整体规划、分期开发的策略，布局统一采用多层及高层的均衡小体量建筑，以满足孵化型企业的使用及经济考量需求，打造符合地域发展的“工业上楼”。突出东西向广场轴线与南北中央景观轴线，配合穿插交错的道路网络以及空中智慧连廊，营造多层次的交通空间，确保了多元业态独立且连通，形成极具吸引力的产业社区，为使用者带来丰富独特的现代化产业空间体验。建筑立面采用金属壁板和铝板作为实体面，配合玻璃幕墙，形成了具有雕塑感的立面形象。

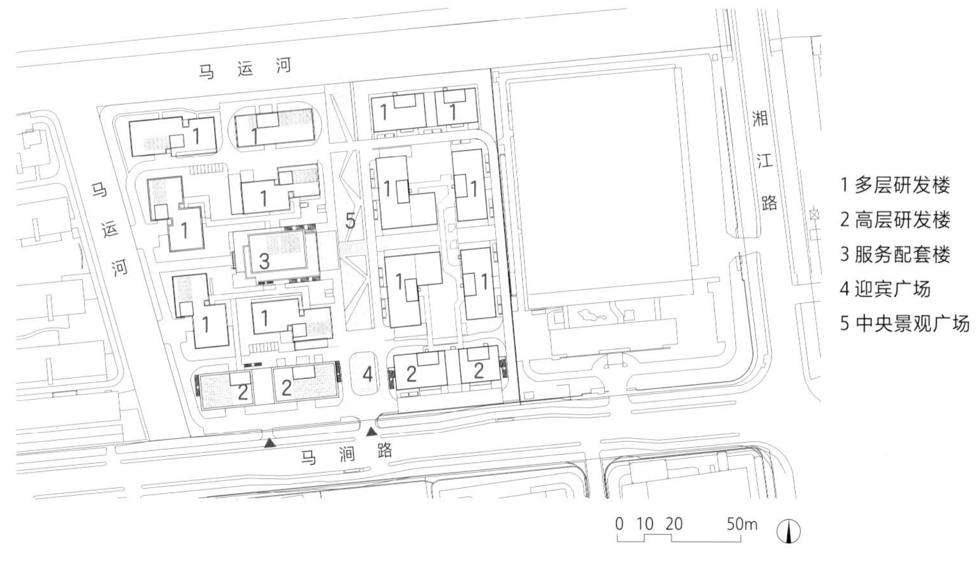

总平面

金唯智基因组研究和基因技术应用实验楼

GENEWIZ GENOME RESEARCH AND GENE TECHNOLOGY APPLICAITION LABORATORY BUILDING

项目类型：科技研发
项目地点：江苏苏州
用地面积：24630m²
建筑面积：65000m²
设计时间：2018年
竣工时间：2021年（一期）

江苏省2018年全过程工程咨询试点项目
第九届江苏省勘察设计行业信息模型（BIM）应用大赛一等奖

项目位于苏州工业园区科教创新区，是金唯智生物科技有限公司的苏州总部办公实验楼、研发布局的重要承载平台和技术产业化转化平台，对于金唯智生物科技有限公司完善区域布局、拓展专业研究领域具有重要意义。

作为科教创新区重要城市节点，项目采用简洁的体块叠加、流畅的曲面线条、现代的色彩和材质搭配。设计并不是简单地将功能集中，而是充分利用地块内部的20m宽城市绿化带，将工作、生活、学习、休闲融为有机整体，形成多层次空间体验，呼应金唯智作为一家有关生命科学的企业形象内核，从而实现生态学、环境学和建筑学的完美结合。

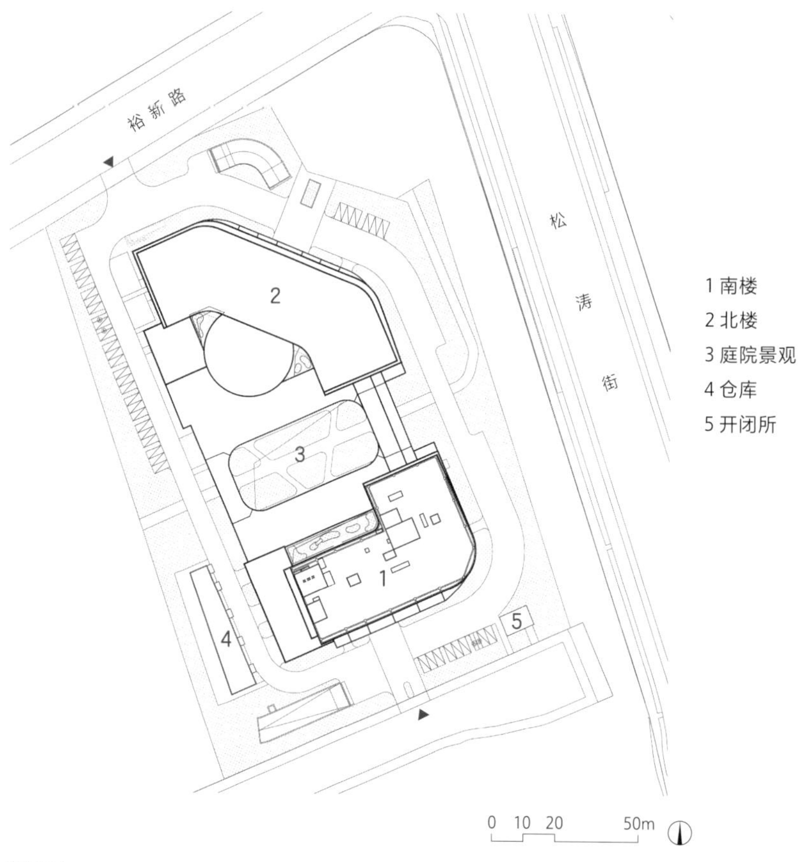

总平面

AZENTA

AZENTA

AZENTA
LIFE SCIENCES

骊住科技新工厂项目

LIXIL TECHNOLOGY NEW FACTORY

项目类型：科研建筑　　**项目地点**：江苏苏州
用地面积：33000m²　　**建筑面积**：46000m²
设计时间：2020年　　**竣工时间**：2022年

项目位于苏州市高新区，作为世界五百强骊住集团苏州总部，融合了研发办公、行政接待以及生产功能。设计以生产所需的工艺流线、空间尺度为主导，在满足业主所需的生产工艺前提下，梳理出其他附属配套功能。在用地条件紧张的情况下尝试不同计容方式的布局组合，以满足不同生产流线所需的不同建筑面积尺寸和建筑空间高度，体现了设计主导的EPC模式的优势。

建筑形象上，考虑到工程造价与企业形象的平衡，设计将VIP入口门厅、生产办公用房布置在厂房北侧，外立面采用玻璃幕墙及穿孔板双层表皮的做法，面向城市界面营造最优企业形象；兼顾造价，其余部位采用真石漆喷涂。建筑主要界面以错落布置的穿孔铝板彰显企业科技感，通过孔洞率的变化营造韵律感。在沉稳克制的色调基础上，用骊住建材生产的橘色陶管点缀入口形象。

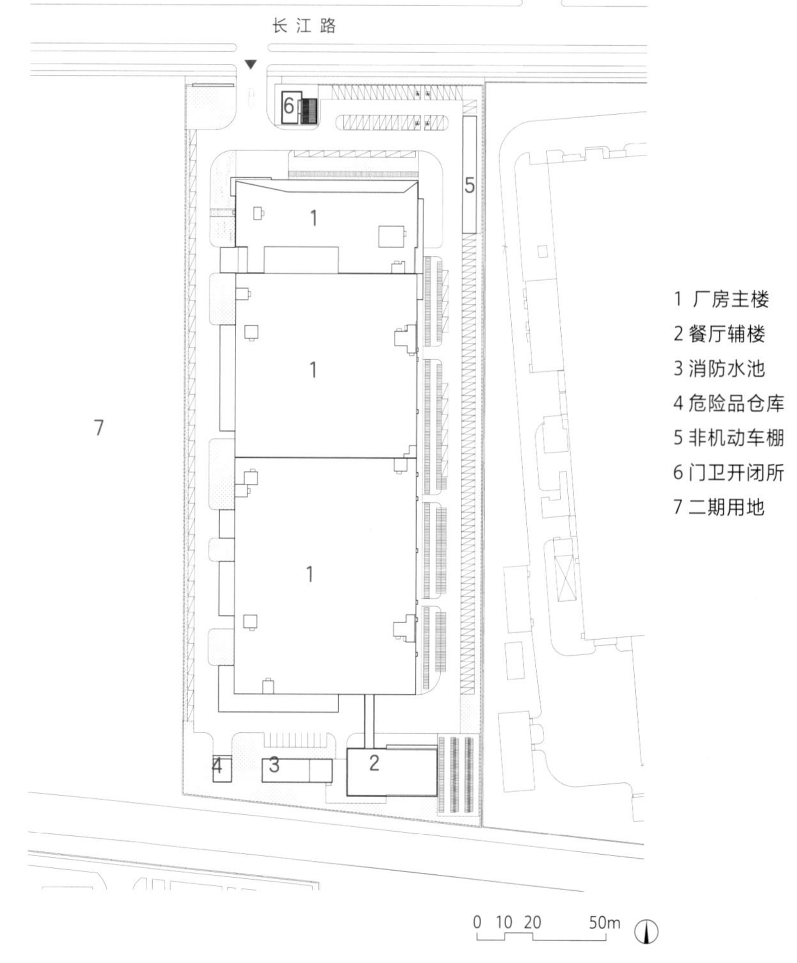

总平面

d Living
骊住建材（苏州）有限公司
LIXIL LINK TO GOOD LIVING

LIXIL Link to Good Living

大兆瓦风机新园区项目

ENVISION GROUP NEW PARK PROJECT

项目类型：科技研发　　**项目地点**：江苏无锡
用地面积：40002m²　　**建筑面积**：44014m²
设计时间：2013年　　**竣工时间**：2015年

2020年度江苏省第十九届优秀工程设计一等奖
2020年度江苏省城乡建设系统优秀勘察设计一等奖

项目位于江苏省江阴市的临港新城低碳产业园内。该项目是由风能发电功能核心构件的生产用房、行政接待、研发办公、员工餐厅等功能区块组合而成。

设计充分考虑到工程造价与企业形象的平衡，通过简洁的体量和干净明快的立面处理手法，塑造优雅而干净的现代厂房形象。设计将VIP入口门厅、生产办公用房布置在厂房西侧，南侧为大型机组构件出入口。面向城市界面的综合楼采用玻璃幕墙及穿孔板双层表皮的做法，面向厂区内部的非主要立面采用彩钢板、真石漆喷涂，以及局部穿孔板，有效兼顾了造价和效果。

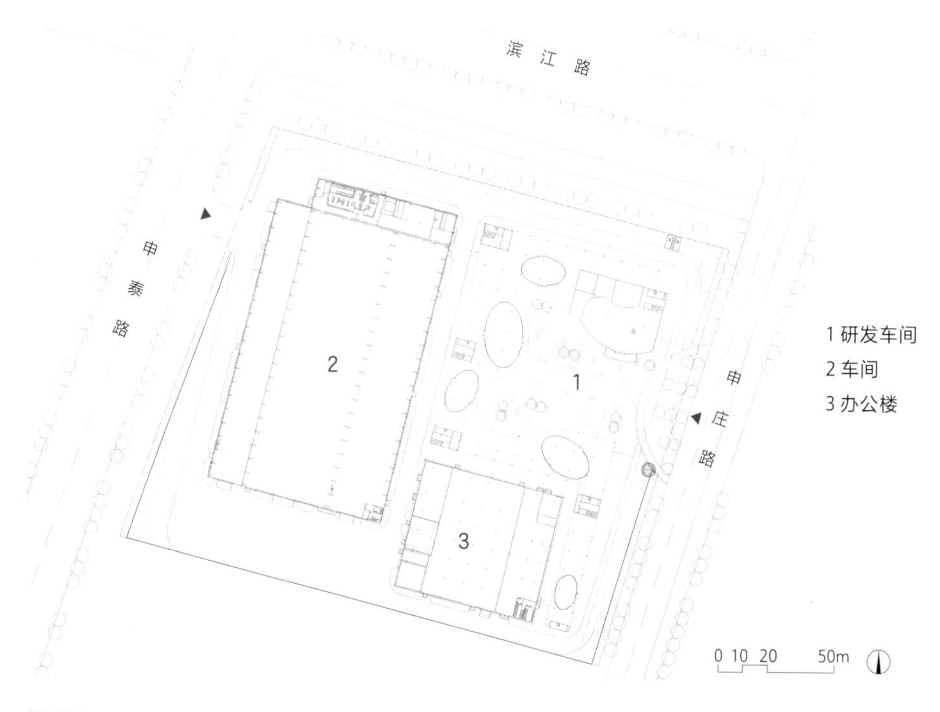

总平面

苏州纳米城

NANOPOLIS SUZHOU

项目类型：产业园区　　**项目地点**：江苏苏州

用地面积：1000000m^2（规划）　　**建筑面积**：1540000m^2（规划）

设计时间：2010年　　**竣工时间**：在建

合作单位：德国海茵建筑设计公司

2015年度全国优秀工程勘察设计奖三等奖
2020年度中国建筑学会建筑设计奖工业三等奖
2015年度第五届“华彩奖”银奖
2013年度江苏省第十六届优秀工程勘察设计二等奖
2013年度江苏省城乡建设系统优秀勘察设计二等奖

项目位于苏州工业园区独墅湖科教创新区，北临金鸡湖大道，西临中环东线，南接独墅湖大道。从2010年开发至今，已成为世界上最大的纳米产业园区。

苏州纳米城旨在聚焦微纳制造技术、清洁能源技术、纳米生物技术和纳米新材料等领域，集聚重大研发机构、国际组织、纳米技术平台、成长型规模型企业，形成具有完备创新与产业化功能的综合产业园区。纳米城总体规划由中心两块产业核心区向四周辐射扩展，地块最外边缘的体块将整个用地围合起来，形成一个类城市的空间结构，设计逻辑清晰，功能分区明确，道路系统完善。

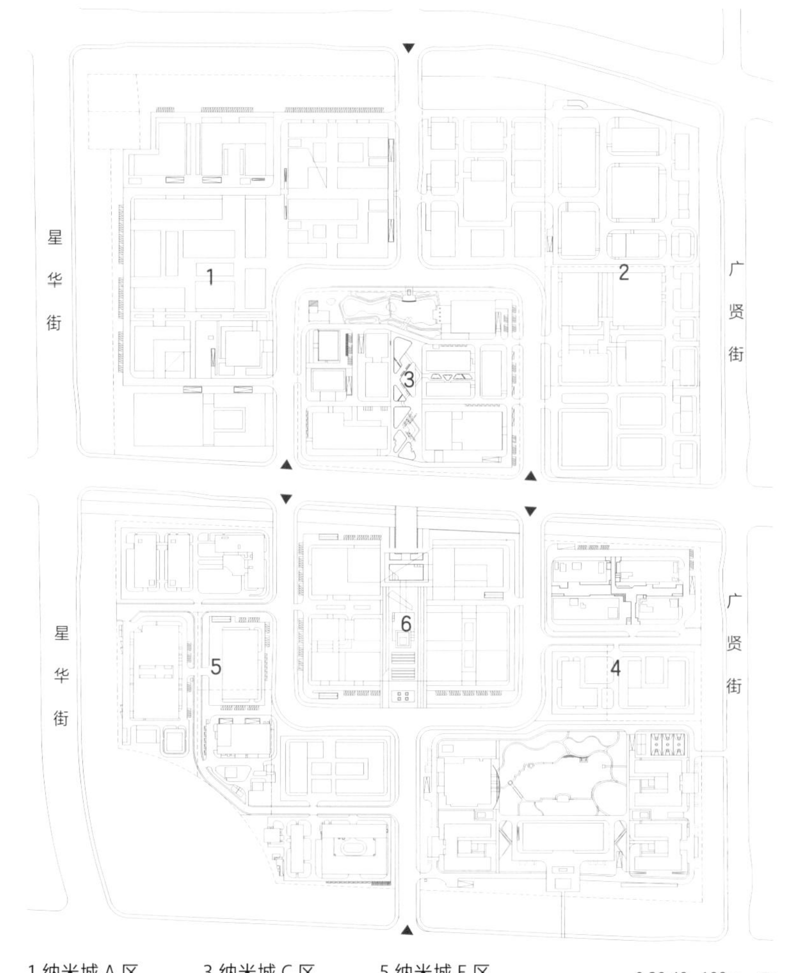

1 纳米城 A 区　3 纳米城 C 区　5 纳米城 E 区
2 纳米城 B 区　4 纳米城 D 区　6 纳米城 F 区

总平面

宝时得中国总部一期

POSITEC CHINA HEADQUARTERS PHASE Ⅰ

项目类型：科技研发　　**项目地点**：江苏苏州
用地面积：41928m^2　　**建筑面积**：33304m^2
设计时间：2011年　　**竣工时间**：2015年
合作单位：伍德佳帕塔设计咨询（上海）有限公司

2015年度全国绿色建筑创新奖三等奖
2017年度全国优秀工程勘察设计行业奖绿建二等奖
2017年度全国优秀工程勘察设计行业奖智能化三等奖
2017年度全国优秀工程勘察设计行业奖三等奖
2018年度中国建筑学会建筑设计奖给排专业三等奖
2015年度第五届“华彩奖”铜奖
2016年度中国建筑学会建筑工程优秀设计奖电气三等奖
2017年度中勘协园林景观分会“计成奖”三等奖
2016年度江苏省城乡建设系统优秀勘察设计二等奖
2016年度江苏省第十七届优秀工程设计三等奖
2019年度江苏省第十八届优秀工程设计景观三等奖
2017年度江苏省优秀工程勘察设计行业奖暖通一等奖
2017年度江苏省优秀工程勘察设计行业奖绿建二等奖
2018年度“江苏省绿色建筑创新奖”二等奖
2017年度江苏省城乡建设系统优秀勘察设计景观二等奖
2016年度江苏省优秀工程勘察设计行业奖电气三等奖
2016年度江苏省优秀工程勘察设计行业奖智能化二等奖

项目位于苏州工业园区东旺路、金堰路交汇的东北侧。建筑南立面由大面积的光伏玻璃组成，光伏玻璃既能为整栋大楼带来清洁能源，又是内部办公空间的遮阳构件。北侧外墙由数十个面积为6m^{2}的玻璃钢采光圆洞组成，与北部外墙之间有2m的间隔，形成自然的双层呼吸幕墙。建筑内部设置了多个导风中庭，利用温度差形成的风压，来加强建筑内部的自然通风。设计团队从可持续场址、水效率、能源与大气、材料与资源、室内环境质量、创新设计六大方面进行可持续设计及项目全过程管理。项目秉承绿色、低碳的开发理念，实现了中国绿色三星级建筑认证和美国LEED-NC铂金级认证。

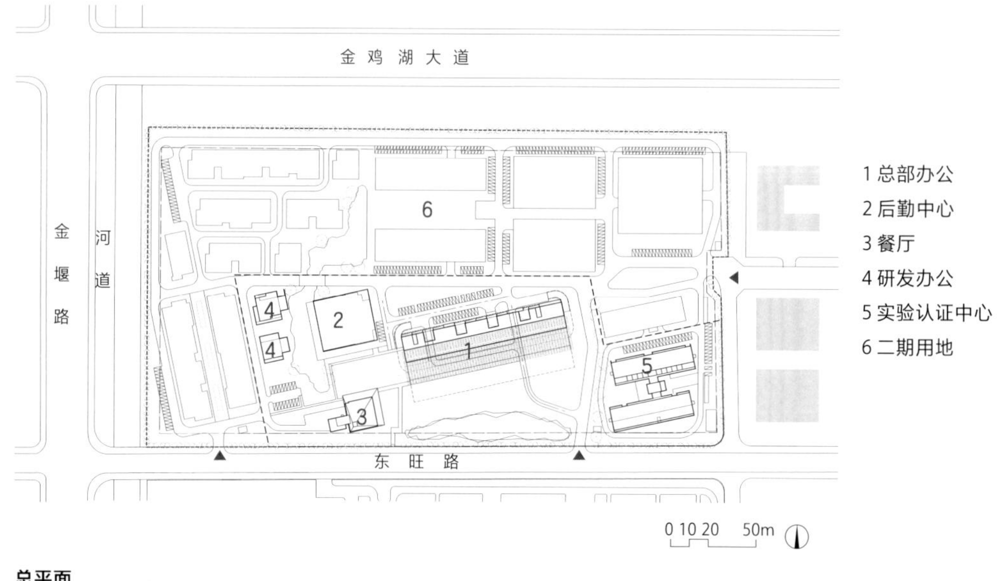

总平面

POSITEC

伯乐蒂森焊接技术苏州厂房

BOHLER WELDING GROUP WORKSHOP IN SUZHOU

项目类型：科技研发　　**项目地点**：江苏苏州
用地面积：43711m²　　**建筑面积**：24700m²
设计时间：2003年　　**竣工时间**：2006年
合作单位：奥地利VISA建筑师事务所

2008年度江苏省第十三届优秀工程设计三等奖
2008年度江苏省城乡建设系统优秀勘察设计三等奖

苏州工业园区为中国和新加坡政府合作开发的国家生态工业示范园区，伯乐蒂森焊接技术（苏州）有限公司位于工业示范园区东南角的四区。

设计力求能突破标准厂房规矩有余、变化不足的方盒子形态，利用生产工艺对建筑空间的长、宽、高三维要求的多样化，遵循“形式追随功能”的理念，创造出现代气息浓厚、体型变化丰富的空间形态。建筑布局根据流线的需要，在办公楼和生产区间设置一条跨度30m的空中钢结构玻璃天桥，轻巧透明，成为整体建筑的一大亮点。立面设计利用长方体的穿插，辅以梯形、圆形及三角形的点缀，构成一组极具雕塑感、体量感的建筑。色彩以红、黄、蓝三原色为主要色调，使建筑更加整体有序且富有现代气息。

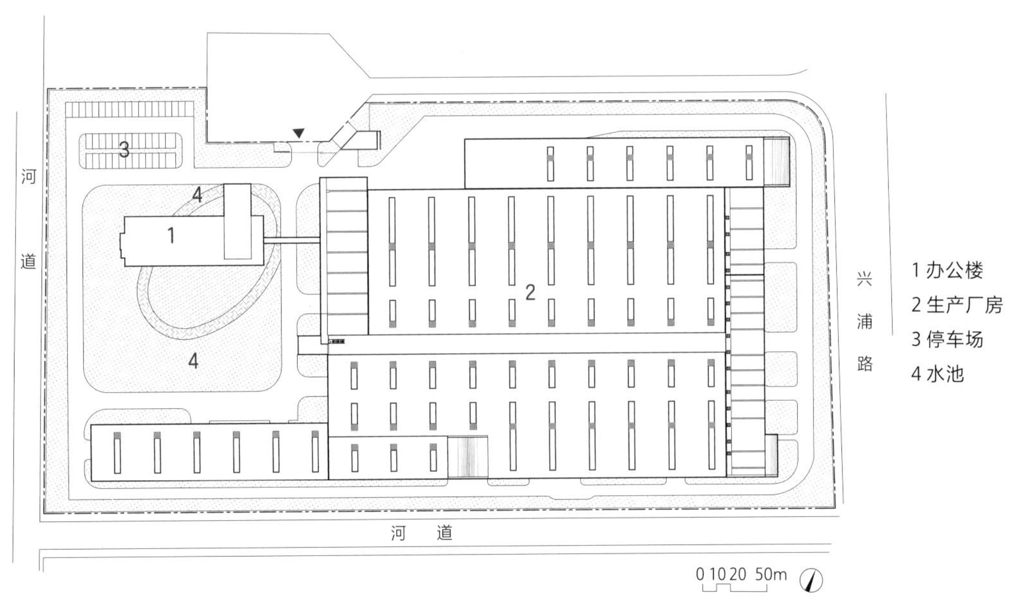

总平面

智创山谷总部园区

ZHICHUANG VALLEY HEADQUARTERS PARK

项目类型：总部园区　　**项目地点**：江苏苏州
用地面积：113937m²　　**建筑面积**：563602m²
设计时间：2022年　　**竣工时间**：在建

项目位于苏州相城经开区，承接相城区“一区十业百园千企”战略规划，推动互联网、先进材料、生物医药、文化产业与经济社会融合发展，建立人才与产业汇集的创新平台，从国家层面打造产业科技示范基地。

基地周边缺乏良好的外部景观环境，因此设计通过围合型建筑布局，力求在园区内部形成一个安静舒适、多维共享的景观体系。设计以莫比乌斯环作为方案的概念原型，通过“以建筑为峰，以庭院为谷”的设计手法，将绿色生态景观渗透其中，营造舒适的空间体验。项目定位总部园区，集企业总部、创新研发、科技孵化、配套公寓、酒店于一体，力争成为相城经济开发区智创企业产城融合的标杆。

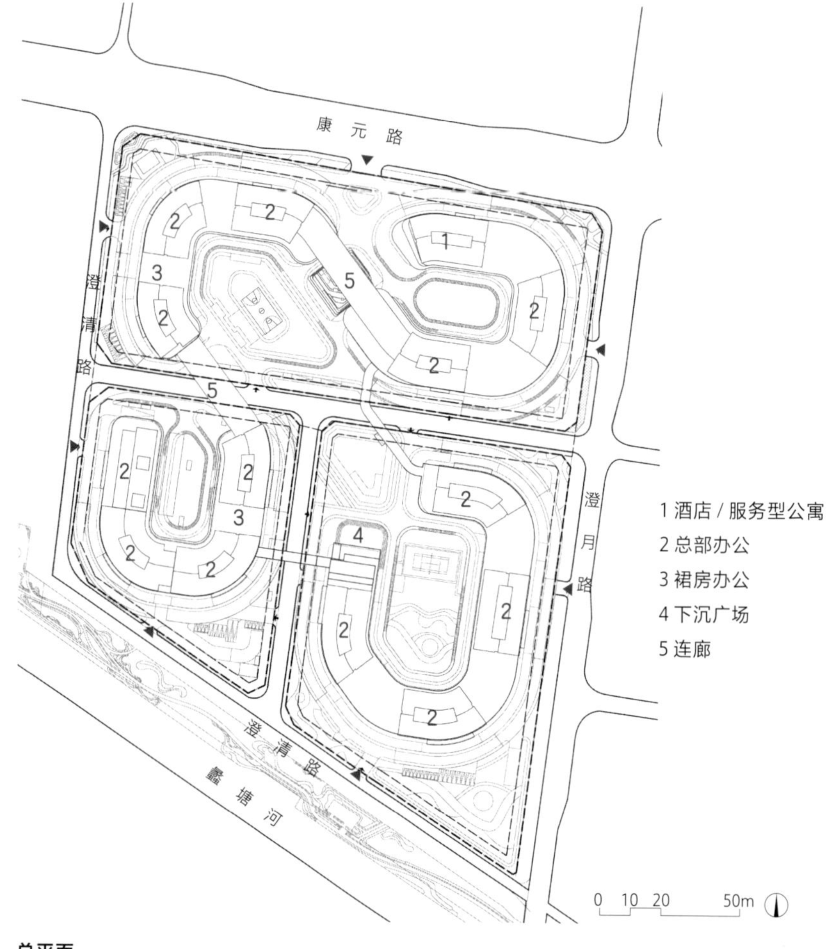

总平面

嘉兴微创园

JIAXING MICROPORT PARK

项目类型：总部园区　　**项目地点**：浙江嘉兴

用地面积：93685m²　　**建筑面积**：280185m²

设计时间：2021年　　**竣工时间**：在建

设计以企业的动脉支架产品为灵感，将企业精神贯穿其中，以人性化的“公园式”办公为设计理念基准，最终将形式转变成建筑语言，生成围绕中央公园的由主脉轴线串联而成的总体布局结构。

相较于传统厂房的机械式排布方式，设计因地制宜，设计了一条贯穿整个厂房区域的共享平台，使所有的建筑与人都能完全无障碍地联系与交流。弧形共享大轴能够激活场所精神，将所有建筑凝聚在一起，形成天然的向心力。中心的接待功能及活动中心成为内向型景观的一部分，也是整体视线的焦点。

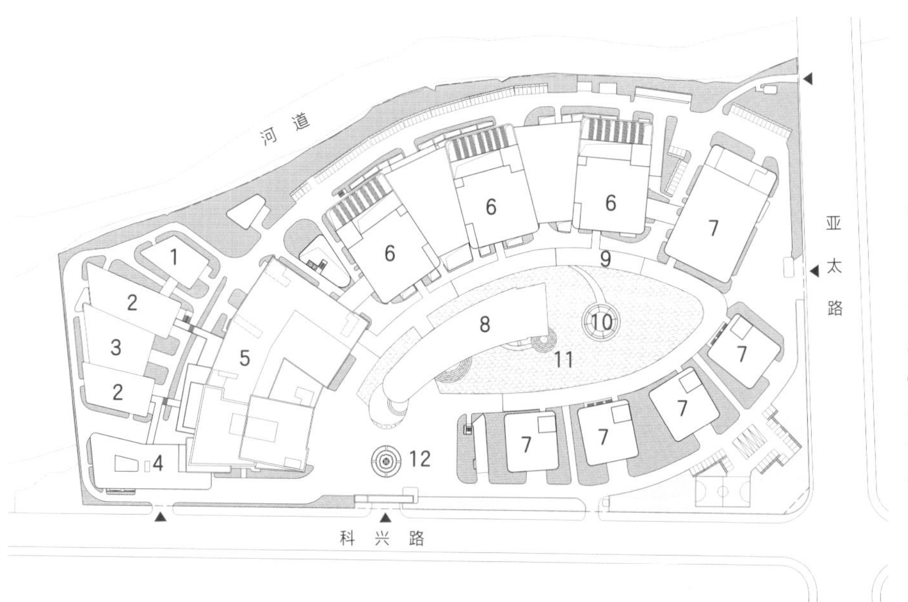

1 VIP餐厅（已建）
2 员工宿舍（已建）
3 员工餐厅（已建）
4 员工宿舍
5 研发办公楼
6 生产用房
7 生产用房（已建）
8 接待中心
9 连廊
10 员工活动室
11 中央水景
12 入口广场

总平面

09

既有建筑改造与乡村振兴

RECONSTRUCTION OF EXISTING BUILDING AND RURAL VITALIZATION

人民路综合整治提升工程
RENMIN ROAD COMPREHENSIVE IMPROVEMENT PROJECT
苏州市政府大楼
SUZHOU MUNICIPAL GOVERMENT BUILDING
苏州会议中心改造提升
UPGRADING OF SUZHOU CONVENTION CENTER
明城墙沿线旧城出新项目
RENEWAL PROJECT ALONG THE MING CITY WALL
树山村改造提升
SHUSHAN VILLAGE TRANSFORMATION AND UPGRADING
阳澄湖镇消泾村二亩塘特色田园乡村
YANGCHENGHU TOWN XIAOJING VILLAGE ERMU POND
CHARACTERISTIC RURAL VILLAGE

人民路综合整治提升工程

RENMIN ROAD COMPREHENSIVE IMPROVEMENT PROJECT

项目类型：城市更新　　**项目地点**：江苏苏州

用地面积：429000m²　　**道路长度**：5.45km

设计时间：2013年　　**竣工时间**：2017年

2020年度中国建筑学会建筑设计奖城市设计三等奖

人民路贯穿苏州古城，自北向南串联起火车站、北寺塔、观前商圈、沧浪亭、南门商圈等重要节点，是伴随苏州逐步生长的城市脊梁，也是展示苏州人文历史、旅游生态的重要窗口。改造中根据人民路的历史内涵以及城市职能将人民路定位为“交通轴线”“景观大道”“文化长廊”。

工程包括人民路全段以及沿线15个重点地块，项目由多专业全方位一体化设计，充分挖掘现代“苏式风格”的精髓，搭建可持续发展框架，采用智能化、数字化技术，充分发挥资源优势，将文化元素运用到立面改造、各类公共空间以及市政设施中，展现古城韵味，唤起古城历史记忆。针对各个重点地块进行详细设计，与古城整体肌理及建筑风貌相协调，以保证古城风貌的整体性和连续性。

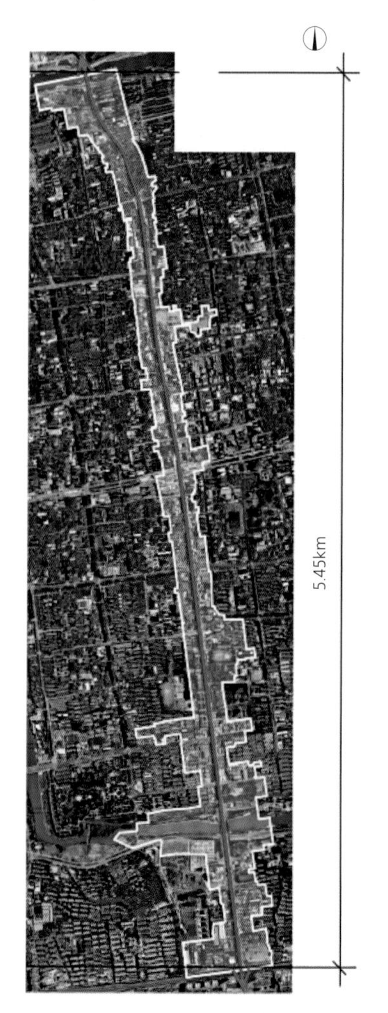

规划范围

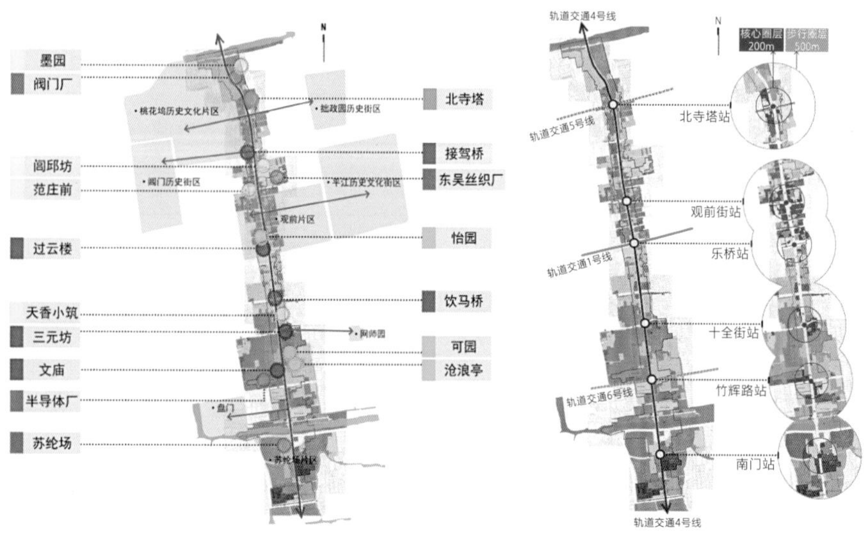

历史人文资源分布

轨道交通站点布局

苏州市政府大楼

SUZHOU MUNICIPAL GOVERMENT BUILDING

项目类型：城市更新　　**设计时间**：2018年
项目地点：江苏苏州　　**竣工时间**：2020年
建筑面积：35453m^2

2022年度第二十届优秀工程设计二等奖
2022年度省城乡建设系统优秀勘察设计二等奖

项目位于姑苏区三香路苏州市政府大院内，建于1996年，属于一类高层办公建筑，已使用二十余年，存在建筑漏水、设备设施老化、故障率高以及原空间无法满足目前实际办公需求等问题。

设计以节约、实用、绿色低碳和适宜办公为出发点，对不适用、不完善的地方进行整修改造，对结构进行整体加固；外形风貌基本保持不变；内部格局按照改造后的功能要求重新设计，在原建筑层高只有3.5m的情况下，以管线优化方式将净高做到2.4m；机电系统结合现状，按照新规范重新进行设计；地下增设通道连接南广场车库直接进入室内；整体按照绿色建筑标准进行改造设计，达到绿色二星级的标准。

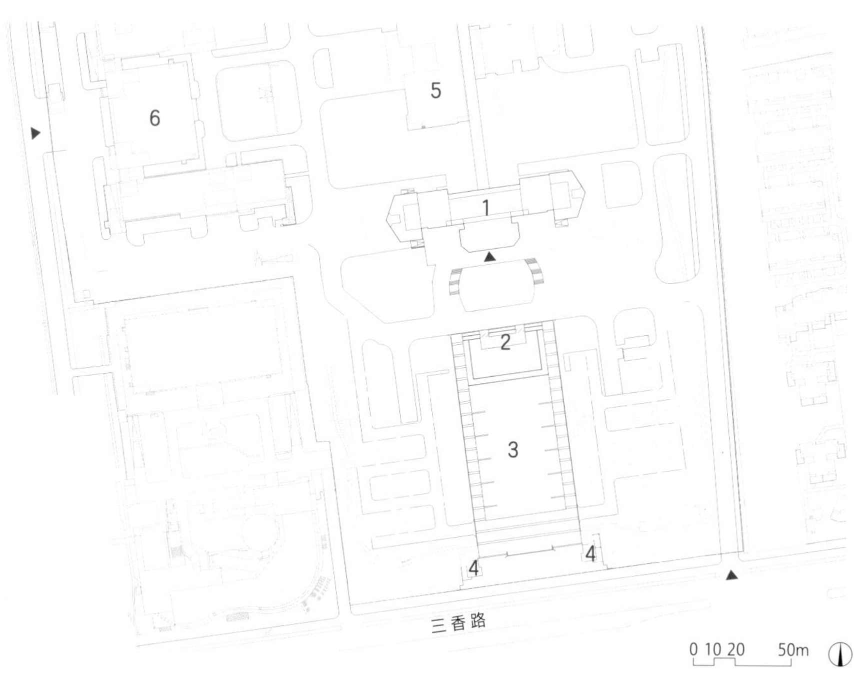

1 行政中心 5 号楼
2 升旗台
3 南广场
4 门卫
5 现存已建 6 号楼
6 现存 10、11、12 号楼

总平面

苏州会议中心改造提升

UPGRADING OF SUZHOU CONVENTION CENTER

项目类型： 城市更新　　**项目地点：** 江苏苏州
用地面积： 43773m²　　**建筑面积：** 87720m²
设计时间： 2019年　　**竣工时间：** 2022年

项目位于姑苏区，南临道前街，西靠养育巷。该项目不仅是苏州市重要会议召开的场所，同时还负责国家和省部级领导人的住宿安排。

项目改造过程面临两个主要的难题。一方面需要考虑如何保留现有建筑和原有场地的印象，以及新老建筑之间的过渡；另一方面，要融合会议场所所需的庄重空间氛围与苏州地域特色。

设计首先对建筑的结构状况、历史价值以及对新功能的适应性等因素进行评估，通过合理的建筑保留和改造，在保持历史记忆的同时，提升会议中心的现代化设施和功能，通过在建筑风格、外观设计、室内空间连通等方面的设计，确保新老建筑之间的过渡自然而流畅。整体建筑风格将江南传统建筑语汇提炼整合，简单明快的同时不失苏州特色。庭院空间考虑运用传统苏州园林的元素和设计手法，如水景、假山、回廊等，为会议中心营造独特的氛围，使现代化的会议功能与苏州的文化底蕴相融合，提供与众不同的体验。

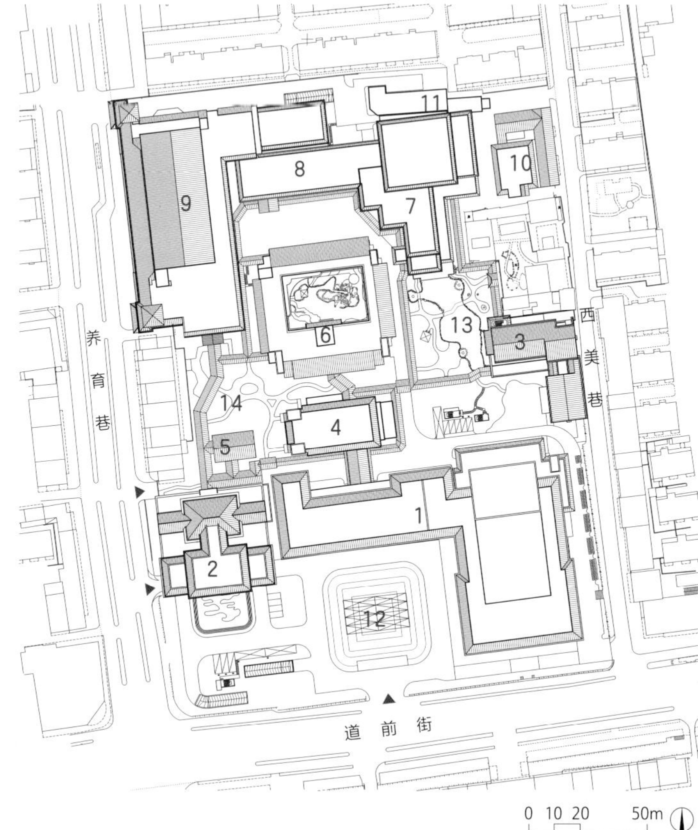

总平面

明城墙沿线旧城出新项目

RENEWAL PROJECT ALONG THE MING CITY WALL

项目类型：城市更新　　**项目地点**：江苏南京

用地面积：78188m²　　**建筑面积**：240650m²

设计时间：2016年　　**竣工时间**：2018年

通过对明城墙沿线老旧住区的有机更新和城墙周边景观环境的塑造，激活了明城墙沿线社区的活力与功能改善了区域交通连通了城市休闲的节点，提升了城市形象。住区改造通过十大项及基础类、完善类、提升类三个层次进行指引，引导老旧小区改造从局部走向系统、从综合整治走向有机更新。不仅提升老旧小区建筑质量、基础设施、公服配套和环境品质，营造安全、健康、便利、人性化的生活空间，同时整合老旧小区周边资源，加强内外空间连接，与宜居街区建设相衔接，促进明城墙片区整体品质提升。

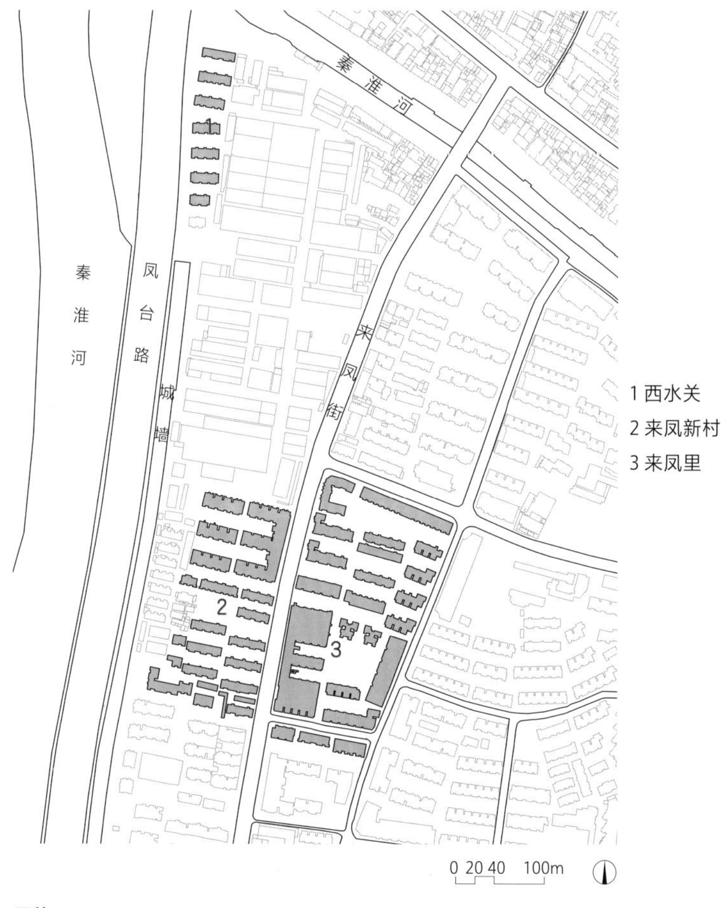

区位

改造前（左）后（右）对比

改造前（左）后（右）对比

树山村改造提升

SHUSHAN VILLAGE TRANSFORMATION AND UPGRADING

项目类型：乡村振兴　　**项目地点**：江苏苏州

用地面积：1.5km^2　　**建筑面积**：103058m^2

设计时间：2017年　　**竣工时间**：2019年

2022年度金拱奖——乡村振兴规划设计金奖

2022年度江苏省第二十届优秀工程设计村镇三等奖

2022年度江苏省第二十届优秀工程设计景观三等奖

2021年度江苏省城乡建设系统优秀勘察设计村镇三等奖

2021年度江苏省城乡建设系统优秀勘察设计景观三等奖

树山村改造工程尊重村内固有的历史文化特色及景观特色，尊重原住民生活习惯及生活方式，重视村内的山水资源，充分保留并发扬村庄的原有印记。村内建筑沿袭江南传统建筑沉静淡雅、粉墙黛瓦的特色；利用原生态的本地材料作为营建要素，深入挖掘村内文化及历史积淀，提炼村内传统精神并以现代手法表达出来，根据各村历史、人文特点形成村内标识、小品或公共空间。注重“诗情画意”的氛围，活用听觉、嗅觉等其他感官共同打造乡村田园风情，在江南传统审美观的基础上，营造既能体现乡村原味、又能融入现代需求的村落群。

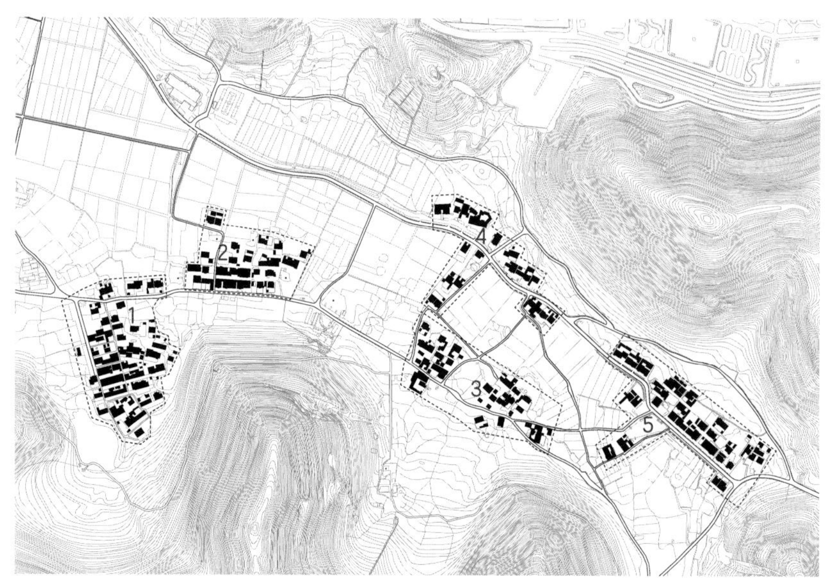

1 戈家坞　2 唐家坞　3 虎窠里南　4 虎窠里北　5 金芝岭

0 20 40 100m

区位

改造前（左）后（右）对比图

改造前（左）后（右）对比图

阳澄湖镇消泾村二亩塘特色田园乡村

YANGCHENGHU TOWN XIAOJING VILLAGE ERMU POND CHARACTERISTIC RURAL VILLAGE

项目类型：乡村振兴　　**设计时间**：2018年

项目地点：江苏苏州　　**竣工时间**：2020年

用地面积：30750m^2

项目设计依托现状独有的“阳澄湖大闸蟹文化”的特色文化背景，以“特色产业”“特色生态”“特色文化”为核心，充分打造“阳澄蟹乡”——江南特色蟹文化水乡风貌，塑造质朴自然的渔家生活体验。消泾村二亩塘特色田园乡村毗邻阳澄湖北岸，外部以清水廊道与阳澄湖相接，是阳澄湖镇发展生态观光农业最重要的示范村。村庄内依存大片蟹塘风貌，可谓是村塘相簇、枕河人家。

设计尊重自然原风貌，打造独有的生态景观格调：村、田、塘融为一体，蟹塘和村宅相依相生，展现风景优美、环境幽静的江南水乡生态。项目充分依托现状水乡鱼塘肌理进行空间营造，创客集市区、净水示范塘区、生态田园区、蟹乐园区四大板块紧密相连，让人们充分体验江南水乡的生态意蕴。设计采用低介入手法，场地地形肌理尽可能予以保留修缮；场地绿化遵循保留现状绿色大基底的原则，多以乡土草本植物及乡村树种进行打造。依托场地乡村风貌，景观设计充分融合大闸蟹文化、乡村水文化、船文化，进行独特的场景表达。

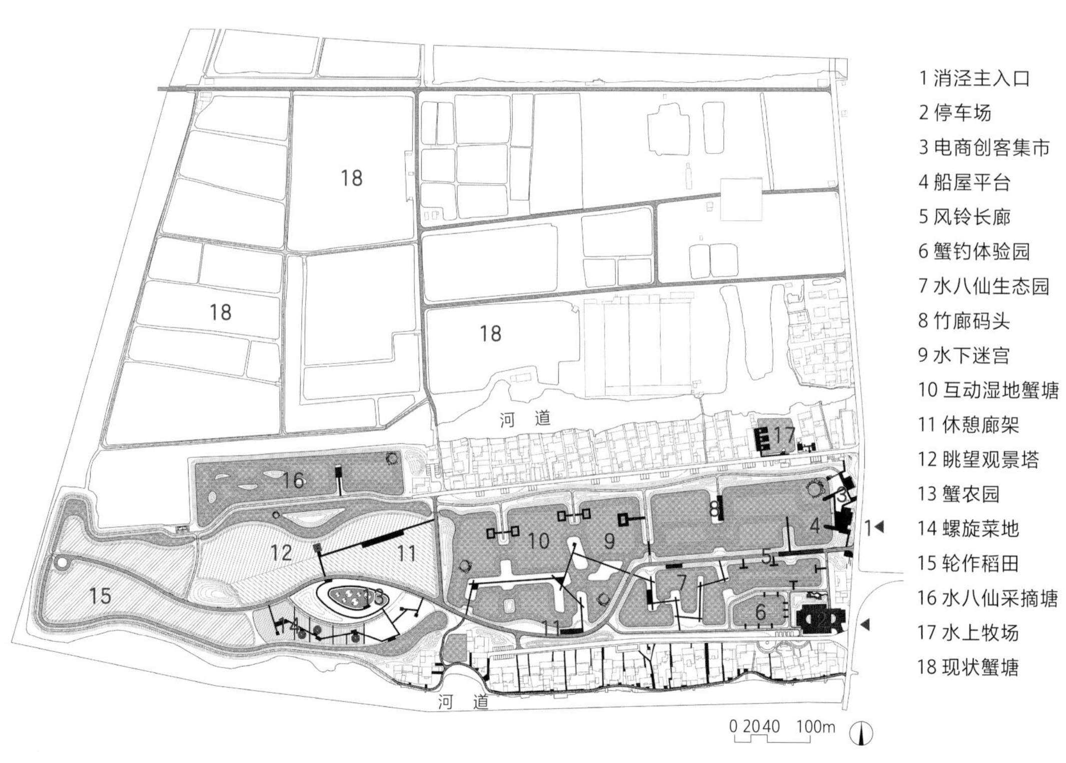

总平面

10

TOD 与基础设施

RAIL TRANSIT AND UNDERGROUND SPACE

星海生活广场
XINGHAI CENTRAL STATION
城市轨道交通车辆段上盖物业综合开发
COMPREHENSIVE DEVELOPMENT OF PROPERTY ON
THE TOP OF URBAN RAIL TRANSIT DEPOT
木里公交换乘枢纽
MULI BUS TRANSFER HUB
杵山交通枢纽
CHUSHAN TRAFFIC HUB
东太湖防汛物资仓库
EAST TAIHU LAKE FLOOD PREVENTION WAREHOUSE

星海生活广场

XINGHAI CENTRAL STATION

项目类型：地下空间　　**项目地点**：江苏苏州
用地面积：14046m²　　**建筑面积**：52273m²
设计时间：2008年　　**竣工时间**：2011年

2012年度江苏省第十五届优秀工程设计三等奖
2012年度江苏省城乡建设系统优秀勘察设计三等奖

项目位于苏州工业园区湖西CBD核心区域星海街以西、苏华路南北两侧，是苏州市首个TOD项目，同时也是首个与苏州轨道交通1号线无缝对接的全地下商业综合体项目。

项目地下功能为商业和地下停车场，地上为城市公共绿地，项目地下空间与周边的环球188、铁狮门苏悦广场、恒宇广场、信托大厦、建行、农行等大厦都有着地下连通道相连，在CBD内引入情景生活的空间形式和消费观念，为CBD区域带来都市休闲环境，舒缓城市节奏。

项目是苏州首个轨道商业综合体，2012年4月正式运营。项目借助轨道交通的“线”与地下一体化的“面”构成商业综合体规划新模式。这座与轨道交通1号线同步设计、同步施工、同步运营的商业综合体，首开苏州系统开发地下空间的先河。项目既和轨道交通星海广场站无缝对接，又是集餐饮、休闲、娱乐、购物于一体的轨道商业综合体；既是商业化运作的城市消费场所，又是具备民生服务功能的公共配套设施，实现了社会效益与商业效益的完美融合，是园区城市化建设与现代服务业发展的一次成功创新。

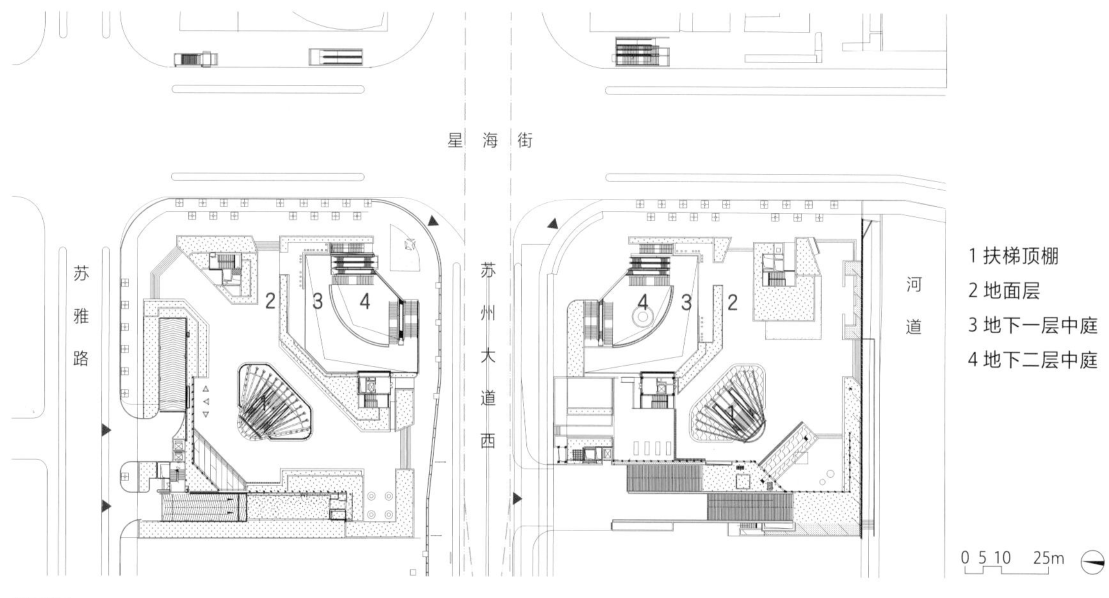

总平面

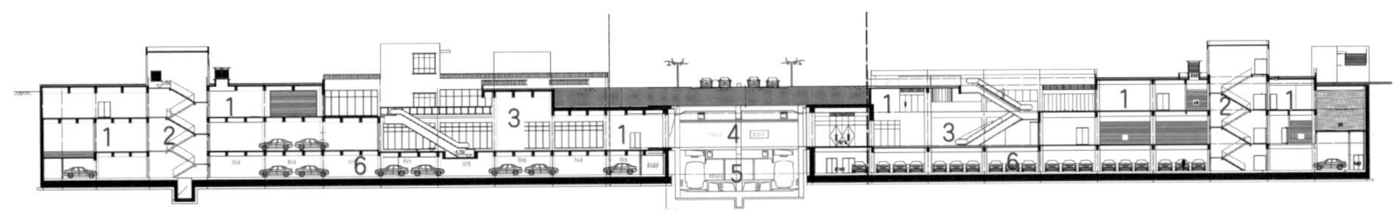

1 商业区
2 垂直交通
3 扶梯
4 地铁非付费区
5 地下轻轨区域
6 地下车库

剖面

城市轨道交通车辆段上盖物业综合开发

COMPREHENSIVE DEVELOPMENT OF PROPERTY ON THE TOP OF URBAN RAIL TRANSIT DEPOT

党的二十大报告提出“构建优势互补、高质量发展的区域经济布局和国土空间体系”。这对轨道交通以及其TOD开发具有极强的战略性指导意义。城市轨道交通车辆段上盖物业综合开发是探索轨道交通建设市场化道路、提升城市功能品质的重要实践方式。系统结构性谋划以及强化轨道与土地利用的联合开发，加快打造具有标杆性的TOD项目，已成为轨道交通高质量发展的时代新命题。

公司近年来完成了国内首个全球WELL健康建筑认证的轨道交通车辆段上盖综合开发社区，江苏省首个交付的车辆基地上盖开发项目，以及正在实施苏州迄今最大规模上盖项目。结合实践经验，形成的《地铁车辆段上盖建筑设计关键技术研究与应用》课题先后获得中国建筑学会科技进步奖、江苏省建设优秀科技成果、江苏省交通运输学会科技奖等，并出版专著《轨道交通车辆基地上盖结构关键技术》1部。

1 苏州轨道交通 2 号线太平车辆段上盖开发项目
2 苏州轨道交通 5 号线胥口车辆段上盖开发项目
3 苏州轨道交通 7 号线天鹅荡车辆段上盖开发项目
4 无锡地铁 4 号线具区路车辆段综合开发项目
5 无锡地铁 5 号线工程新阳路车辆段上盖项目
6 南通轨道交通 2 号线幸福车辆段上盖项目

木里公交换乘枢纽

MULI BUS TRANSFER HUB

项目类型：交通基础设施建筑　　**项目地点**：江苏苏州
用地面积：15271m²　　**建筑面积**：16152m²
设计时间：2018年　　**竣工时间**：2022年

项目位于苏州市吴中区，西侧临近东太湖，用地东、南面紧邻苏州市轨道交通4号线支线天鹅荡停车场上盖项目，其上盖平台采用都市景观的方式营造绿色新城和智慧新城。

项目借鉴了“上盖”设计的概念，将公交和环卫两个单位的停车场及其配套用房上下“叠加”在一起，安置进了原本只能容纳公交停车枢纽使用的地块，实现了场地的叠合利用。为减少作业干扰，设计对公交和环卫车辆的动线进行分流，使其相对独立。基于共享的理念，对两家单位的配套用房进行整合，以优化空间组合。停车场质朴裸露的混凝土构件被金属网包裹，与办公建筑精致的玻璃和铝板幕墙形成对比，相得益彰。

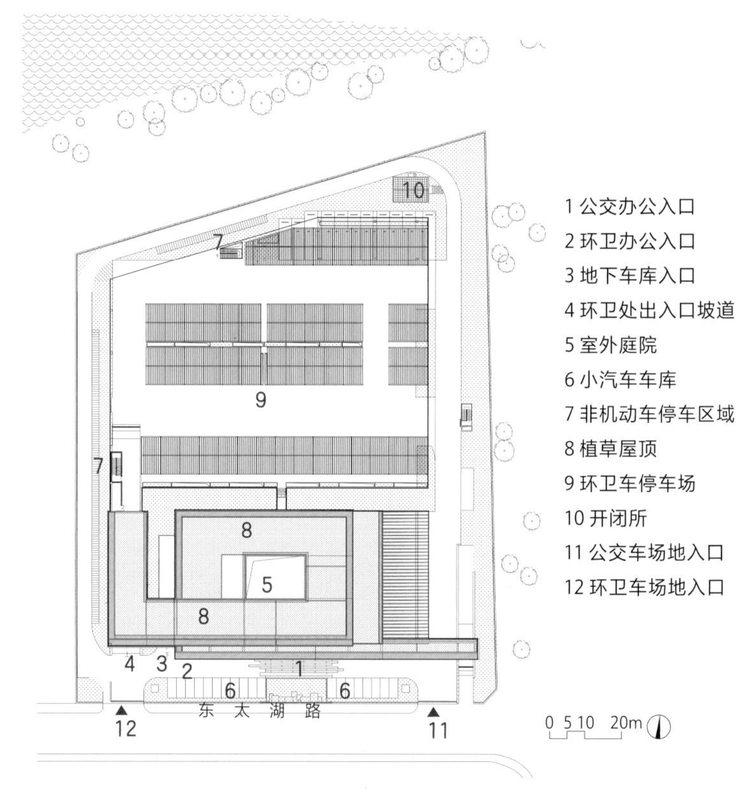

总平面

杵山交通枢纽

CHUSHAN TRAFFIC HUB

项目类型：交通基础设施建筑　　**项目地点**：江苏苏州
用地面积：40741m^2　　**建筑面积**：34253m^2
设计时间：2019年　　**竣工时间**：2022年

2022年度江苏省第二十届优秀工程设计二等奖
2022年度江苏省第二十届优秀工程设计园林绿化工程设计三等奖
2022年度江苏省城乡建设系统优秀勘察设计二等奖
2022年度江苏省城乡建设系统优秀勘察设计园林绿化工程设计二等奖

项目与南太湖毗邻，以“一心、二环、三片区”为规划格局，以游客中心为内核，构建自然山水骑行及生态体验骑行两大环线，打造了换乘区、雨水花园区、交通区三大区域。

设计根据现场场地现状及使用要求，以认知自然、体验自然为宗旨，设置与城市多连接线的交通体系，创造体验丰富的功能场地，加强游人与景观空间的互动，并追求建筑与环境的和谐共生。建筑选材呼应当地特色，以木、竹为主，建筑造型以水滴飘带为原型，在古典的选材中融入新兴的曲面技术，将建筑塑造为环境中的一景，提升区域的整体美感，体现和谐性与艺术性。

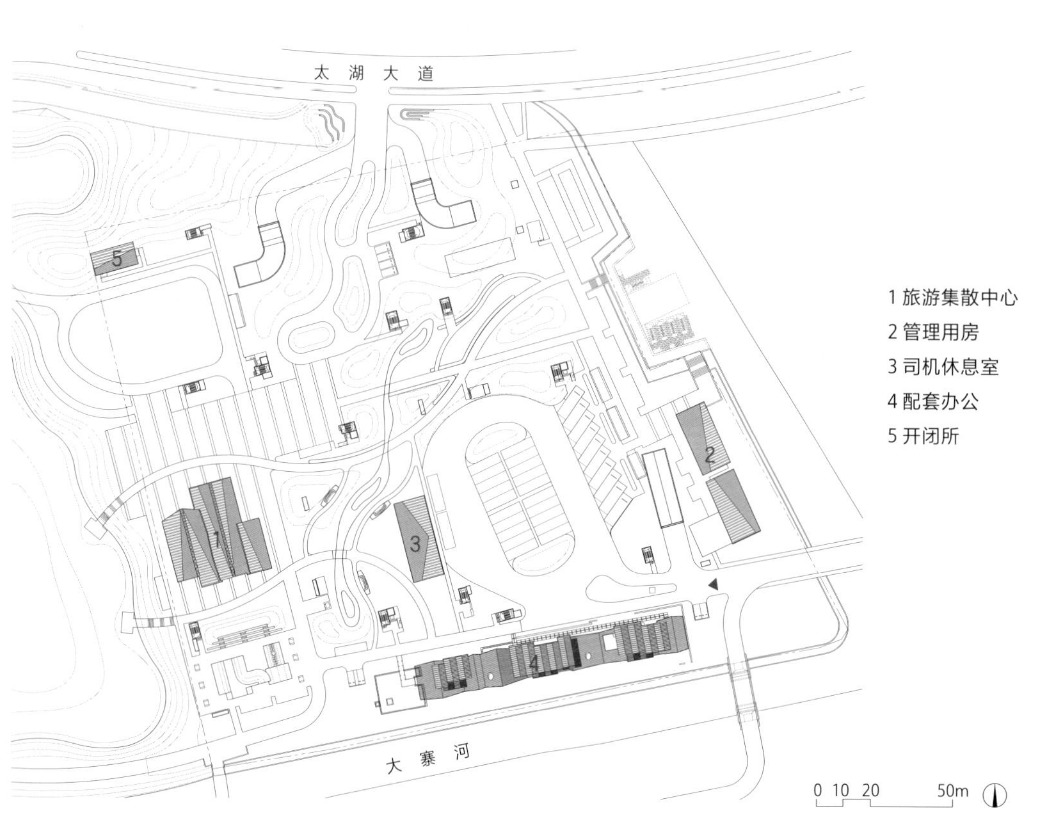

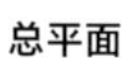

总平面

东太湖防汛物资仓库

EAST TAIHU LAKE FLOOD PREVENTION WAREHOUSE

项目类型：市政基础设施建筑　　**项目地点**：江苏苏州
用地面积：4154m^2　　**建筑面积**：1379m^2
设计时间：2014年　　**竣工时间**：2016年

2020年度中国建筑学会建筑设计奖三等奖
2020年度江苏省第十九届优秀工程设计一等奖
2019年江苏省城乡建设系统优秀勘察设计一等奖
2018年度江苏省土木建筑学会第十二届“建筑创作奖”一等奖
2020年度第七届江苏省勘察设计行业建筑信息模型（BIM）应用大赛最佳BIM建筑设计创意奖
2020 ASIA DESIGN PRIZE WINNER
2021 German Design Award Special Mention
2020 ICONIC AWARDS: Innovative Architecture WINNER
ARCHITECTURE MASTERPRIZE 2020 ARCHITECTURAL DESIGN AWARD WINNER

项目是一个用于太湖防汛的物资储备仓库，体量并不大，所以把如何体现建筑功能以及保持场地的原生态作为设计的主旨。仓库采用斜坡屋顶，绿植由湖边草坪延伸至屋顶覆土，自然地把仓库“掩埋”在环境之中。仓库投入使用后，屋顶绿植和周边景观经过一段时间的生长，使建筑与周边环境充分融合，湖水、草地、绿树和鲜花，把仓库隐藏在滨水环境中。建筑的整个结构体系是内聚性的，在维持足够的功能空间的同时，更好地“消隐”在环境里。设计让仓库“消极”地被掩埋，时间让建筑“积极”地与环境一道共同生长，一个人为、一个自发，让这个仓库最终成为了环境的一部分。

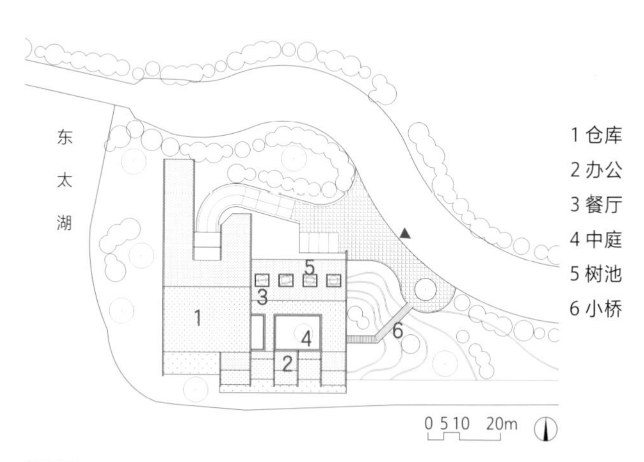

总平面

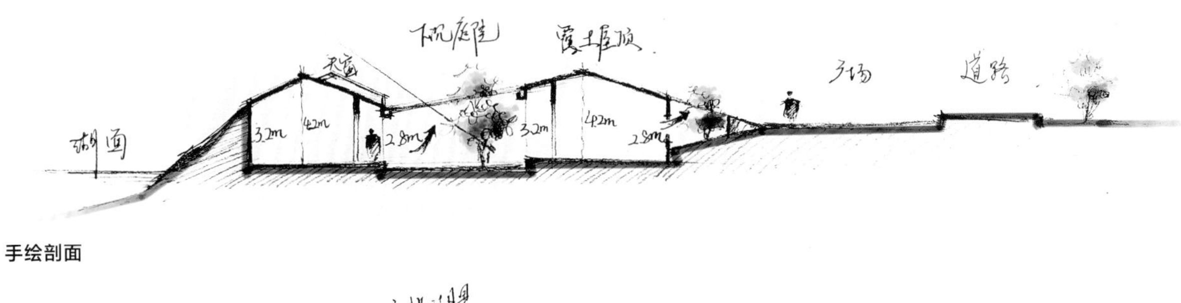
下沉庭院
覆土屋顶
天窗
广场
道路
湖面
3.2m
4.2m
2.8m
3.2m
4.2m
2.8m

手绘剖面

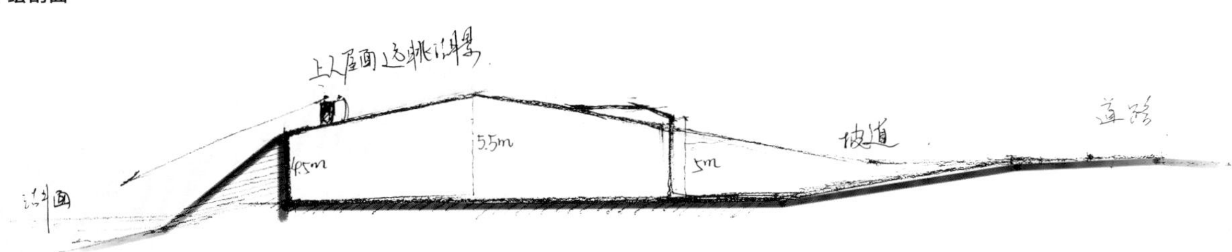
上人屋面远眺湖景
湖面
4.5m
5.5m
5m
坡道
道路

手绘立面

11

城市规划与设计

URBAN PLANNING AND DESIGN

赤峰高新技术产业开发区一体化规划设计
INTEGRATED PLANNING AND DESIGN OF CHIFENG HIGH TECH INDUSTRIAL DEVELOPMENT ZONE
苏州科技城核心区 TOD 综合开发城市设计
URBAN DESIGN OF TOD COMPREHENSIVE DEVELOPMENT IN THE CORE AREA OF SUZHOU SCIENCE AND TECHNOLOGY CITY
无锡南站综合发展区启动区概念规划及城市设计
CONCEPTUAL PLANNING AND URBAN DESIGN OF THE LAUNCHING AREA OF WUXI SOUTH RAILWAY STATION COMPREHENSIVE DEVELOPMENT ZONE
常熟昆承片区概念规划及城市设计
CONCEPTUAL PLANNING AND URBAN DESIGN OF CHANGSHU KUNCHENG DISTRICT
太湖科学城战略规划与概念性城市设计
STRATEGIC PLANNING AND CONCEPTUAL URBAN DESIGN OF TAIHU SCIENCE CITY
常熟西泾岸历史文化街区保护规划
CONSERVATION PLANNING OF XIJINGAN HISTORICAL AND CULTURAL BLOCK
无锡滨湖区马山街道阖闾社区村庄规划
VILLAGE PLANNING OF HELV COMMUNITT, MASHAN STREET, WUXI BINHU DISTRICT

赤峰高新技术产业开发区一体化规划设计

INTEGRATED PLANNING AND DESIGN OF CHIFENG HIGH TECH INDUSTRIAL DEVELOPMENT ZONE

项目类型：城市规划与设计　　**设计时间**：2016年

项目地点：内蒙古自治区赤峰市　　**项目规模**：129km^2

规划的总体发展目标为“经济、社会、生态协调发展的现代化千亿级园区、国家级高新技术产业开发区”。规划引入“有机共生、产城融合”的理念，合理利用资源，营建可持续发展的开发区，形成“多组团、多廊道”的空间结构。将赤峰高新区打造成为我国重要的高新技术产业基地，巩固赤峰东北工业走廊的核心地位，挺起赤峰工业脊梁。

空间结构规划

0 2000 5000m
1000

土地利用规划

苏州科技城核心区TOD综合开发城市设计

URBAN DESIGN OF TOD COMPREHENSIVE DEVELOPMENT IN THE CORE AREA OF SUZHOU SCIENCE AND TECHNOLOGY CITY

项目类型： 城市规划与设计　　**设计时间：** 2020年

项目地点： 江苏苏州　　**项目规模：** 2km^2

合作单位： 清华大学建筑设计研究院有限公司

上海市隧道工程轨道交通设计研究院

设计提出“环太湖中央科创区、山水人文生活典范区、苏州西部综合交通枢纽”三大定位，构建科技城核心区多维枢纽功能圈层模型。设计以“太湖星链，姑苏叠境”为愿景，以TOD综合发展为核心理念，重点研究站点周边多维空间关系，倡导建筑复合化、交通低碳化、服务网络化、活力全域化，优化“山一水一城”空间格局，对外展示生态、人文、创新、活力交织交融的繁华图景，打造“山水长卷，姑苏盛境”的城市空间意向。

无锡南站综合发展区启动区概念规划及城市设计

CONCEPTUAL PLANNING AND URBAN DESIGN OF THE LAUNCHING AREA OF WUXI SOUTH RAILWAY STATION COMPREHENSIVE DEVELOPMENT ZONE

项目类型：城市规划与设计　　**设计时间**：2020年
项目地点：江苏无锡　　**项目规模**：$12km^2$
合作单位：北京清华同衡规划设计研究院有限公司

无锡南站综合发展区启动区概念规划及城市设计竞赛第一名

无锡南站是环太湖科创带和无锡“长江-太湖城市发展轴”的重要交点，规划紧扣“十四五”发展重点，通过深入剖析无锡南站枢纽对周边发展的带动作用，梳理城市发展与太湖生态保护的关系，基于枢纽地区的综合交通规划设计，打造集文化会展、金融商务、总部办公、文化科技、休闲娱乐于一体的城市综合中心。

常熟昆承片区概念规划及城市设计

CONCEPTUAL PLANNING AND URBAN DESIGN OF CHANGSHU KUNCHENG DISTRICT

项目类型：城市规划与设计　　**设计时间**：2021年

项目地点：江苏常熟　　**项目规模**：46km^2

2021年度江苏省土木建筑学会第十五届建筑创作奖一等奖

2022年度江苏省土木建筑学会第十六届“建筑创作奖”城市设计三等奖

规划充分尊重场地优质的生态资源，围绕江河湖浜的江南生境骨架和塘浦圩田的水乡文化底蕴，用生态本位去梳理空间生长逻辑，描绘出一幅物种共生共栖、人民安乐安康、思维交流交响、都市协作协同、产业兴盛兴旺的昆承画卷。

规划总平面

太湖科学城战略规划与概念性城市设计

STRATEGIC PLANNING AND CONCEPTUAL URBAN DESIGN OF TAIHU SCIENCE CITY

项目类型：城市规划与设计　　**设计时间**：2021年
项目地点：江苏苏州　　**项目规模**：77km²
合作单位：清华大学建筑设计研究院有限公司
Tekuma Frenchman

2022年度省土木建筑学会第十六届“建筑创作奖”城市设计一等奖
太湖科学城战略规划与概念性城市设计国际竞赛第二名

规划紧紧围绕“国际一流科学城”这一目标，依托“生态+科技”的思路，力争树立宜居宜业新标杆，为实现“环太湖科学中心”的愿景打好规划基石。设计以水为意向，形成一条贯穿科学城的科创山水走廊，构建生态城市的结构，凸显水乡特色风貌。以科创山水走廊为载体，容纳科研、实验等创新功能。融合科技与文化的水岸空间不但延续了独特的水乡城市肌理，也为科学城提供了高品质的滨水生活。

规划总平面

常熟西泾岸历史文化街区保护规划

CONSERVATION PLANNING OF XIJINGAN HISTORICAL AND CULTURAL BLOCK

项目类型：城市更新
设计时间：2021年
项目地点：江苏常熟
项目规模：175800m²

2023年度苏州市城市更新专项竞赛铜奖

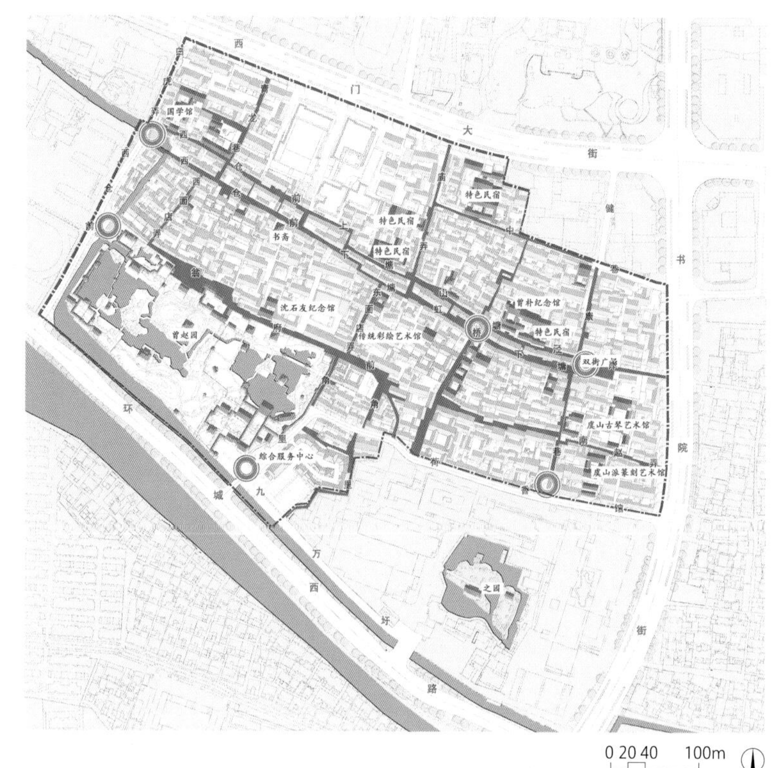

保护规划总平面

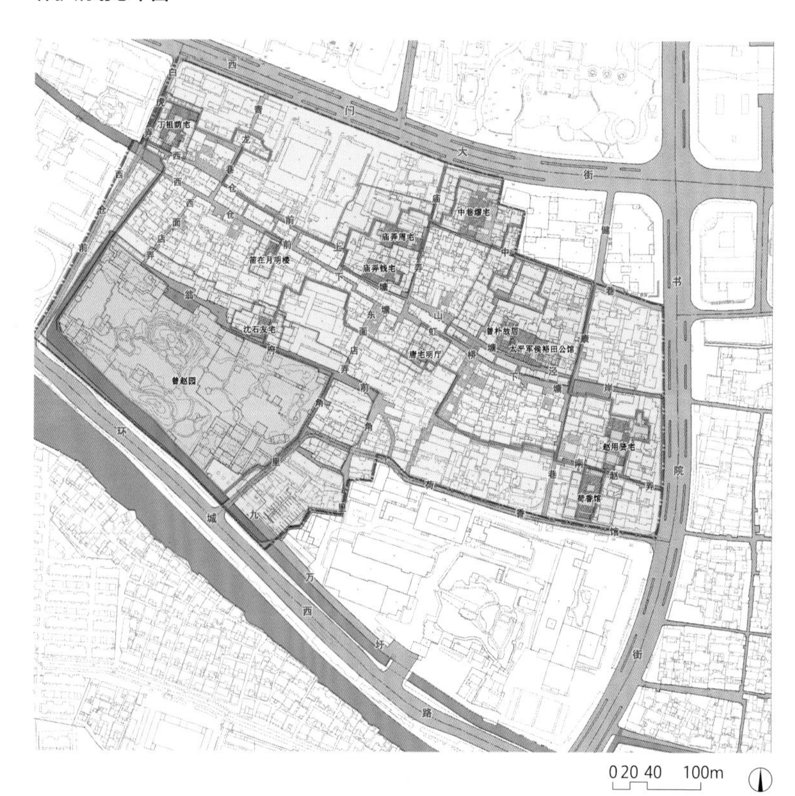

空间格局保护规划

项目规划借助苏州市域古城一体化发展机遇，依托独特区位，围绕文化气韵，结合“特色文化体验、双街活力空间、品质康养生活”3大核心路径，重塑历史质真、文艺之美、风尚至善的新时代“真美善”街区，从而实现西泾岸历史文化街区的保护、传承和复兴。

无锡滨湖区马山街道阖闾社区村庄规划

VILLAGE PLANNING OF HELV COMMUNITT, MASHAN STREET, WUXI BINHU DISTRICT

项目类型：城市更新　　**设计时间**：2020年

项目地点：江苏无锡　　**项目规模**：526hm^2

自然资源部第一批全国国土空间规划（村庄规划）优秀案例

2021年度江苏省优秀国土空间规划二等奖

2020年度省土木建筑学会第三届城市设计专项奖三等奖

2021年度无锡市优秀国土空间规划评选一等奖

规划以乡村振兴为总目标，以“绿水青山就是金山银山”理论为指导，通过对山水人文脉络的梳理，构建生态优先、文化引领、产业振兴等发展路径，形成特有的“阖闾模式”指引村庄发展，助力实现“千年古邑、山水阖闾”的发展愿景。

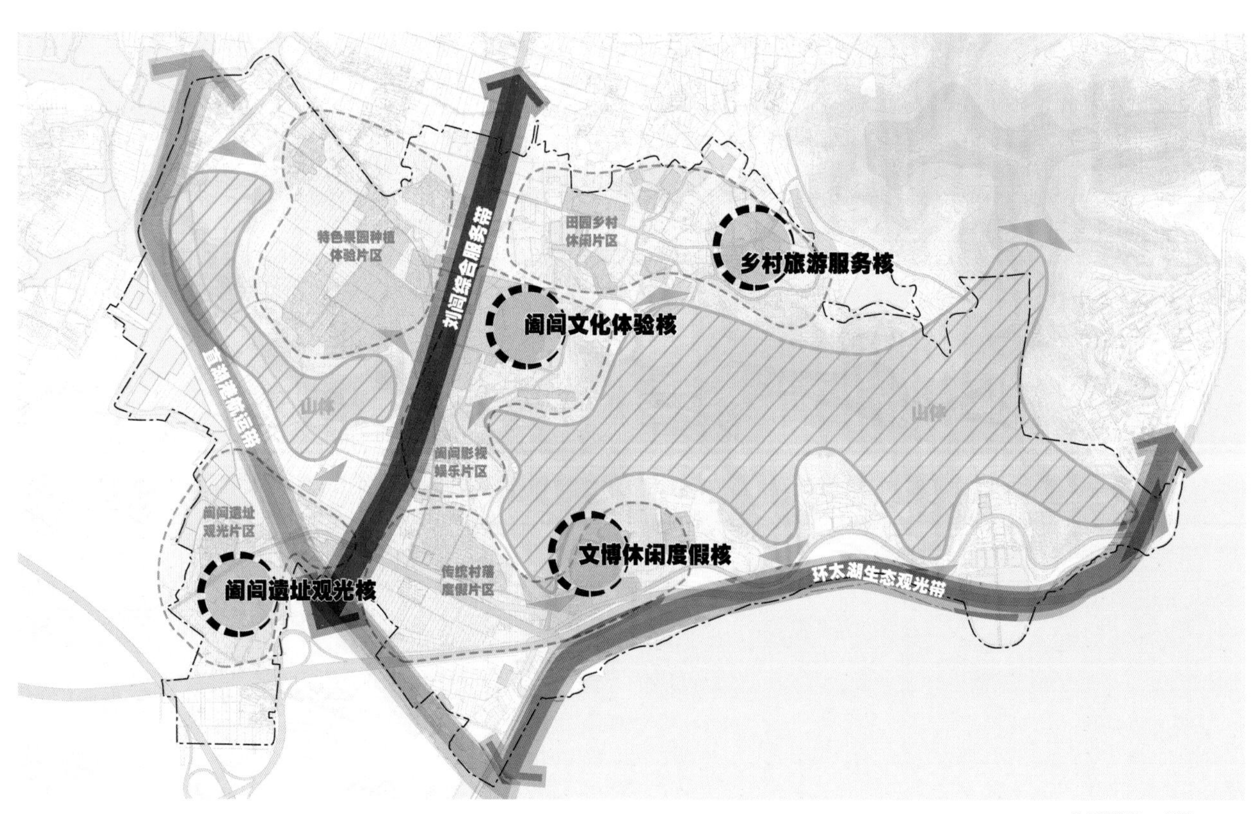

村庄规划结构图

12

生态景观

ECOLOGICAL LANDSCAPE

环元荡生态岸线提升工程

ECOLOGICAL SHORELINE IMPROVEMENT PROJECT AROUND YUANDANG

宁波滨海华侨城示范区延伸段

NINGBO BINGHAI OCT DEMONSTRATION AREA EXTENSION SECTION

四川南部水城禹迹岛公园

YUJI ISLAND PARK SHUICHENG SOUTH SICHUAN

遂宁五彩缤纷路景观公园

SUINING COLORFUL ROAD LANDSCAPE PARK

金鸡湖大道景观改造

LANDSCAPE RENOVATION OF JINJIHU AVENUE

银川艾依河滨水公园

YINCHUAN AYI RIVER WATERFRONT PARK

环元荡生态岸线提升工程

ECOLOGICAL SHORELINE IMPROVEMENT PROJECT AROUND YUANDANG

项目类型：生态景观设计　　**设计时间**：2020年
项目地点：江苏苏州　　**竣工时间**：2021年
项目规模：192000m^2

2022年度第十三届园冶杯专业奖市政园林大奖
2022 ARCHITECTURE MASTERPRIZE BEST OF BEST
2023 DNA Paris Design Awards WINNER

设计意象来自于水乡“鱼群洄游”的形态和水面产生的关系。整体湖岸线及步道设计为流线型。在空间结构上，通过一条一级步道，贯穿、衔接入口三角绿地、杉林氧吧、闲梦云台等区域，同时设置元荡慢行桥，打通苏州与上海边界，助力长三角生态绿色一体化示范区建设。

宁波滨海华侨城示范区延伸段

NINGBO BINGHAI OCT DEMONSTRATION AREA EXTENSION SECTION

项目类型：生态景观设计　　**设计时间**：2021年
项目地点：浙江宁波　　**竣工时间**：2022年
项目规模：14309m^2

2022 MUSE Design Awards GOLD WINNER
2022 DNA Paris Design Awards WINNER
2022年度第二届AHLA亚洲人居景观奖银奖
2022 Global Future Design Awards SILVER

项目位于阳光海湾示范区东侧，设计敬重原生态山海之美，在延续原有示范区景观风格基础上，保留拥有土地记忆的植物与石材，着力以质朴简约的设计语言，让珍稀山海资源成为场地的主角，营造具有山海特色的景观体验与独特记忆。以自然的笔触对话环境，在原生态山海景观中，带来一段回归生活、回归自由、回归热爱的度假旅程。

四川南部水城禹迹岛公园

YUJI ISLAND PARK SHUICHENG SOUTH SICHUAN

项目类型：生态景观设计
项目地点：四川南充
项目规模：1000000m²
设计时间：2019年
竣工时间：2021年

为了营造更完善的生态系统，在规划之初，充分考虑滨江带与城市滨水公园的结合，由内到外构建了“江边滩涂—湿地系统—原生林—人工化园林区域—城市边界”的纵深林线，让自然和城市完美融合。针对地幅阔大绵长的外滩区域，设计通过地形的塑造、植物疏朗关系的调整、密植组团与道路的围合、停驻场地的打造以及特色旱溪的营造，提供了丰富多样的视觉体验，使游客在游览的过程中，既能感受到山河磅礴的壮丽，又能有游园细节上的回味。

“人道我居城市里，我疑身在万山中”。设计以归于自然、融于自然的去风格化手法，平衡河道开发与生态自然的矛盾，营造出历久弥新、生生不息的新型水岸景观空间。春生夏长，秋落冬藏，在禹迹岛公园，人与水以最亲切的形式重新连结，与城市共呼吸，续写着亲水南部新的记忆篇章。

遂宁五彩缤纷路景观公园

SUINING COLORFUL ROAD LANDSCAPE PARK

项目类型：生态景观设计　　**设计时间**：2008年
项目地点：四川遂宁　　**竣工时间**：2012年
项目规模：725000m^2

2015中国“建筑新传媒奖”景观设计大奖
2012中国环境艺术奖—中国最佳环境艺术设计
2011中国人居典范建筑规划设计奖最佳设计方案金奖

设计提出“以优美江岸线为链”的构思，在五彩缤纷路上进行景观的多维度呈现，并进而将之演化成一条灵动的“飘带”。通过在河流原生滩涂中新建自然生态公园，并以此作为回归生态的实践，建筑与景观相互渗透，直至彻底融合，勾勒出一幕极具感染力的城市生态图景。

金鸡湖大道景观改造

LANDSCAPE RENOVATION OF JINJIHU AVENUE

项目类型：道路景观　　**设计时间**：2008年
项目地点：江苏苏州　　**竣工时间**：2009年
项目规模：350000m^2

2011年度全国优秀工程勘察设计行业奖市政公用工程三等奖
2010年度江苏省第十四届优秀工程设计风景园林二等奖
2010年度江苏省城乡建设系统优秀勘察设计风景园林二等奖

设计充分利用道路两侧原有的30m宽城市绿化带，选择恰当的位置布置景观活动场地，将金鸡湖与独墅湖的湖景资源有机地结合进来。全路以双湖景观和森林景观为背景，采用韵律变化的手法点缀人工景观，加大林冠线的起伏变化。道路改造后沿线景观面貌焕然一新，实现了城市空间品质的提升。

银川艾依河滨水公园

YINCHUAN AYI RIVER WATERFRONT PARK

项目类型：生态景观设计　　**设计时间**：2012年
项目地点：宁夏回族自治区银川　　**竣工时间**：2014年
项目规模：192000m^2

2015英国世界建筑新闻奖“滨水设计奖”WAN Waterfront Award
2015首届深圳创意设计七彩奖
2014世界建筑节（World Architecture Festival）景观类入围作品

设计从“融合地域特色、凸显生态低碳、以人为本”三大主题入手，展示多彩银川的魅力与品位。设计立足场地空间的落差，结合银川生态立市、塞上湖城的城市定位，打破绿化隔离的客观存在，缩小城市与自然之间的距离。

13

室内设计

INTERIOR DESIGN

无锡太湖华邑酒店
WUXI TAIHU HUALUXE HOTEL
三亚凯悦嘉轩酒店
HYATT PLACE HOTEL SANYA
中山保利艾美酒店
LE MERIDIEN ZHONGSHAN
三亚万科别墅
SANYA VANKE VILLA
绿盟科技成都新川 5 号
NO.5, XINCHUAN CHENGDU OF NSFOCUS
深圳万科中心室内设计改造
INTERIOR DESIGN RENOVATION OF SHENZHEN VANKE CENTER
恒荣控股办公会所
HENGRONG HOLDING OFFICE CLUB
鸿荣源深圳北站商业中心
HONGRONGYUAN SHENZHEN NORTH RALIWAY STATION BUSINESS CENTER

无锡太湖华邑酒店

WUXI TAIHU HUALUXE HOTEL

项目类型：酒店
项目地点：江苏无锡
项目规模：29980m²
设计时间：2014年
竣工时间：2017年

项目精心提炼当地传统的人文符号、民间材质、中国色彩等，将之转变为设计手段及元素，构筑了一个富含东方美学的场景，平和内敛、朴质风雅，传统中透着现代气息，现代中糅合古典之韵。

三亚凯悦嘉轩酒店

HYATT PLACE HOTEL SANYA

项目类型： 酒店
项目地点： 海南三亚
项目规模： 27055m^2
设计时间： 2015年
竣工时间： 2018年

设计将动感的“波浪”引入空间，在充满变化的异形建筑体内延续并展开海洋絮语，唤醒商旅时间的奇妙遐想。柔美鲜艳的色彩和独特舒适的材质纹理调和出一个个“沉浸式”体验场景，激荡着温润人心的滨海城市之歌。

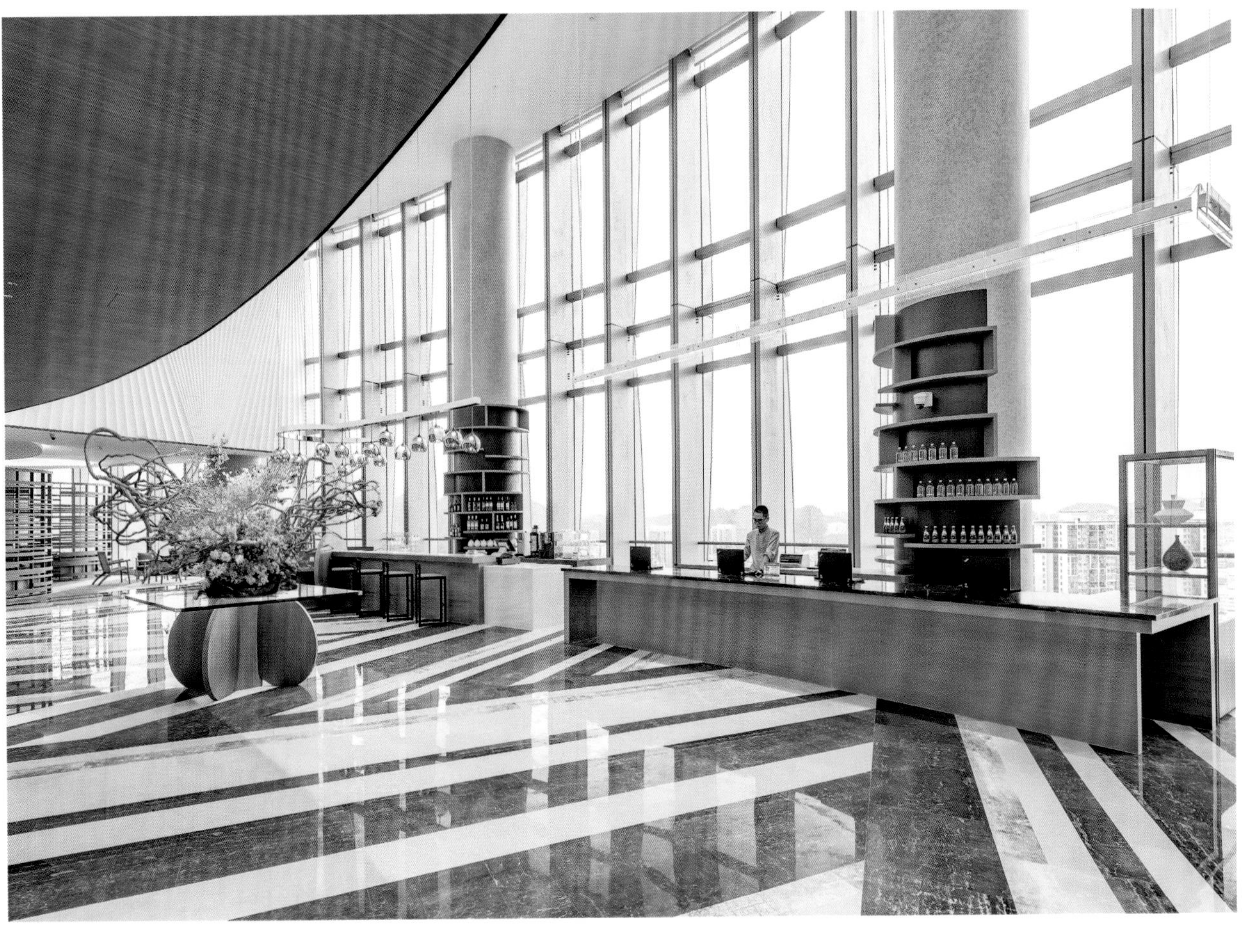

中山保利艾美酒店

LE MERIDIEN ZHONGSHAN

项目类型：酒店
项目地点：广东中山
项目规模：35327m²
设计时间：2015年
竣工时间：2018年

设计从全新的视角发掘古镇文化魅力，以“山”元素为灵感，将酒店品牌文化中的法式优雅注入其中，使空间趣味横生，激发人们的好奇与探索。以香山之韵营造简约独特的法式风情，将当代美学引进这座历史悠久的城市。古香山与新中山的相遇，真切回应着对生活的美好向往，激发着人们对于城市奥秘的永恒探索。

三亚万科别墅

SANYA VANKE VILLA

项目类型：住宅
项目地点：海南三亚
项目规模：745m²
设计时间：2017年
竣工时间：2018年

空间通过个性化的材质统一构建低调的灰白背景，在精巧雅致的度假主题之下，不同色彩和质感的器物有机组合，呈现自然生机与人类智慧的和谐共融。在巧妙的设计下，各个功能空间取得了最佳的采光通风和观景视野效果。挑高的天窗设计、别致的庭院景观和开放式的格局，让互相独立的景致共生交融，形成有趣的空间对景。

绿盟科技成都新川5号

NO.5, XINCHUAN CHENGDU OF NSFOCUS

项目类型：办公
项目地点：四川成都
项目规模：9203m^2
设计时间：2018年
竣工时间：2019年

2020年度第十六届中国国际建筑装饰及设计艺术博览会华鼎奖金奖
2022年度第五届AIIDA美国国际创新设计大奖 装饰 国际创新奖

在了解公司的行业属性后，设计团队提取“盾”作为核心设计元素，并且将这种元素运用到具体设计中，构造出一座“盾之堡垒”。项目设计充分考虑了环保和低碳的要求，在材料规格上遵循建筑的模数进行排列组合，以减少材料的损耗，提高利用率。

深圳万科中心室内设计改造

INTERIOR DESIGN RENOVATION OF SHENZHEN VANKE CENTER

项目类型：办公
项目地点：深圳
项目规模：2500m^2
设计时间：2015年
竣工时间：2016年

伴随万科快速发展带来的人员扩张，以及办公方式的改变，万科总部急需一个集接待、会议、社交为一体的升级空间，此次改造任务意义非同一般。设计展示了一个复合型的新概念办公空间：明快自在的接待区、自然温馨的会议区，在扮演各自重要角色功能的同时，却又制造着互动和惊喜。

恒荣控股办公会所

HENGRONG HOLDING OFFICE CLUB

项目类型：办公
项目地点：四川成都
项目规模：$1730m^2$
设计时间：2021年
竣工时间：2022年

项目位于成都万科大厦内，设计以独特的“巴适”生活美学为灵魂，构建了一处人文与艺术相融、科技与创意辉映的雅趣之所。整体空间分为办公、会所两大功能区，工作与生活、公共与私密，看似矛盾的需求，通过材质与颜色的差异、高度与光线的区分获得边界感，实现不同的空间表达。

鸿荣源深圳北站商业中心

HONGRONGYUAN SHENZHEN NORTH RALIWAY STATION BUSINESS CENTER

项目类型：办公
项目地点：深圳
项目规模：11831m²

设计时间：2019年
竣工时间：2020年

项目位于北站商务中心区核心区内，紧邻高铁核心枢纽深圳北站，是联系商务中心片区与商业文化片区的纽带。基于商务大堂及营销中心双重功能定位，设计以打造开放共享的复合型大堂为出发点，提取并延展“V”形元素作为建筑语汇，摒弃多余的装饰，以理性的美学方式，展现厚重的气度，凸显舒适细腻的人文关怀，树立起北站商务中心区的建筑典范。

14

经典作品

CLASSIC PROJECTS

1953—1990 >>>

新华书店
Xinhua Bookstore

项目类型：商业
项目地点：江苏苏州
竣工时间：1978
Project Type：Commercial
Location：Suzhou, Jiangsu
Completion：1978

苏州火车站（旧）
Suzhou Railway Station

项目类型：交通基础设施
项目地点：江苏苏州
竣工时间：1982
Project Type：Transport infrastructure
Location：Suzhou, Jiangsu
Completion：1982

苏州刺绣研究所接待馆
Reception Hall of Suzhou Embroidery Research Institute

项目类型：服务性建筑
项目地点：江苏苏州
竣工时间：1982
Project Type：Service building
Location：Suzhou, Jiangsu
Completion：1982

苏州彩香住宅小区规划
Suzhou Caixiang Residential District Planning

项目类型：住宅
项目地点：江苏苏州
竣工时间：1984
Project Type：Residence
Location：Suzhou, Jiangsu
Completion：1984

苏州市园外楼饭店
Suzhou Yuanwailou Hotel

项目类型：饭店
项目地点：江苏苏州
竣工时间：1984
Project Type：Hotel
Location：Suzhou, Jiangsu
Completion：1984

苏州市总工会友谊宾馆
Suzhou Labour Union Friendship Hotel

项目类型：宾馆
项目地点：江苏苏州
竣工时间：1988
Project Type：Hotel
Location：Suzhou, Jiangsu
Completion：1985

苏州竹辉饭店
Suzhou Bamboo Grove Hotel

项目类型：酒店
项目地点：江苏苏州
竣工时间：1990
Project Type：Hotel
Location：Suzhou, Jiangsu
Completion：1990

北京东直门公寓（东湖宾馆）
Beijing Dongzhimen Apartment（Donghu Hotel)

项目类型：酒店
项目地点：北京
竣工时间：1990
Project Type：Hotel
Location：Beijing
Completion：1990

1991—2000 >>>

苏州丝绸博物馆
Suzhou Silk Museum

项目类型：博物馆
项目地点：江苏苏州
竣工时间：1991
Project Type：Museum
Location：Suzhou, Jiangsu
Completion：1991

苏州市建筑设计研究院业务楼
SIAD Building

项目类型：办公
项目地点：江苏苏州
竣工时间：1992
Project Type：Office
Location：Suzhou, Jiangsu
Completion：1992

苏州寒山寺弄整修改建
Renovation of Suzhou Hanshan Temple Lane

项目类型：既有建筑改造
项目地点：江苏苏州
竣工时间：1992
Project Type：Reconstruction of existing building
Location：Suzhou, Jiangsu
Completion：1992

苏州革命博物馆
Suzhou Revolutionary Museum

项目类型：博物馆
项目地点：江苏苏州
竣工时间：1993
Project Type：Museum
Location：Suzhou, Jiangsu
Completion：1993

苏州市外事办公室
Suzhou Foreign Affairs Office

项目类型：办公
项目地点：江苏苏州
竣工时间：1994
Project Type：Office
Location：Suzhou, Jiangsu
Completion：1994

苏州万盛大厦
Suzhou Labour Union Friendship Hotel

项目类型：宾馆
项目地点：江苏苏州
竣工时间：1994
Project Type：Hotel
Location：Suzhou, Jiangsu
Completion：1994

苏州国贸中心
Suzhou International Trade Center

项目类型：会展
项目地点：江苏苏州
竣工时间：1994
Project Type：Exhibition
Location：Suzhou, Jiangsu
Completion：1994

苏苑饭店二期客房楼
Suyuan Hotel Phase II Guest Room Building

项目类型：酒店
项目地点：江苏苏州
竣工时间：1994
Project Type：Hotel
Location：Suzhou, Jiangsu
Completion：1994

苏州桐芳巷试点小区
Suzhou Tongfang Lane Pilot Community

项目类型：住宅
项目地点：江苏苏州
竣工时间：1996
Project Type：Residence
Location：Suzhou, Jiangsu
Completion：1996

1991—2000 >>>

雅都大酒店
Aster Hotel

项目类型：酒店
项目地点：江苏苏州
竣工时间：1994
Project Type：Hotel
Location：Suzhou, Jiangsu
Completion：1994

长发商厦
Changfa Commercial Building

项目类型：商业
项目地点：江苏苏州
竣工时间：1995
Project Type：Commercial
Location：Suzhou, Jiangsu
Completion：1995

苏州人民商场
Suzhou People's Mall

项目类型：商业
项目地点：江苏苏州
竣工时间：1997
Project Type：Commercial
Location：Suzhou, Jiangsu
Completion：1997

苏州河东新区综合楼
Suzhou Hedong New Area Complex Building

项目类型：行政办公楼
项目地点：江苏苏州
竣工时间：1996
Project Type：Administrative office building
Location：Suzhou, Jiangsu
Completion：1996

苏州市新区管委会涉外用房
Foreign Related Housing of SND Management Committee

项目类型：行政办公楼
项目地点：江苏苏州
竣工时间：1998
Project Type：Administrative office building
Location：Suzhou, Jiangsu
Completion：1998

中信实业银行苏州市分行大楼
CITIC Industrial Bank Suzhou Branch Building

项目类型：金融办公
项目地点：江苏苏州
竣工时间：1999
Project Type：Financial office
Location：Suzhou, Jiangsu
Completion：1999

中国银行苏州分行干将路综合业务楼
Bank of China Suzhou Branch Ganjiang Road

项目类型：金融办公
项目地点：江苏苏州
竣工时间：1999
Project Type：Financial office
Location：Suzhou, Jiangsu
Completion：1999

工商银行胥门支行业务楼
ICBC Xumen Sub branch

项目类型：金融办公
项目地点：江苏苏州
竣工时间：1996
Project Type：Financial office
Location：Suzhou, Jiangsu
Completion：1996

苏州泰华商城
Suzhou International Trade Center

项目类型：商业
项目地点：江苏苏州
竣工时间：1996
Project Type：Commercial
Location：Suzhou, Jiangsu
Completion：1996

南浩街（北段）低洼区改造
Reconstruction of Low-lying Areas in Nanhao Street

项目类型：既有建筑改造
项目地点：江苏苏州
竣工时间：2000
Project Type：Reconstruction of existing buildings
Location：Suzhou, Jiangsu
Completion：2000

苏州工艺美术职业技术学院老校区
Old Campus of Suzhou Art &Design Technology Institute

项目类型：教育建筑
项目地点：江苏苏州
竣工时间：1996
Project Type：Educational building
Location：Suzhou, Jiangsu
Completion：1996

苏州市新区一中
SND No. 1 Middle School

项目类型：教育建筑
项目地点：江苏苏州
竣工时间：1999
Project Type：Educational building
Location：Suzhou, Jiangsu
Completion：1999

江苏省泰州中学新校区
New Campus of Taizhou Middle School

项目类型：教育建筑
项目地点：江苏泰州
竣工时间：1999
Project Type：Educational building
Location：Taizhou, Jiangsu
Completion：1999

锦华苑高层公寓
Jinhuayuan High Rise Apartment

项目类型：住宅
项目地点：江苏苏州
竣工时间：1998
Project Type：Residence
Location：Suzhou, Jiangsu
Completion：1998

陈云故居暨纪念馆辅助设施
Chen Yun's Former Residence

项目类型：文化建筑
项目地点：上海
竣工时间：2000
Project Type：Cultural architecture
Location：Shanghai
Completion：2000

国巨电子（苏州）有限公司
YAGEO

项目类型：工业厂房
项目地点：江苏苏州
竣工时间：1999
Project Type：Industrial plant
Location：Suzhou, Jiangsu
Completion：1999

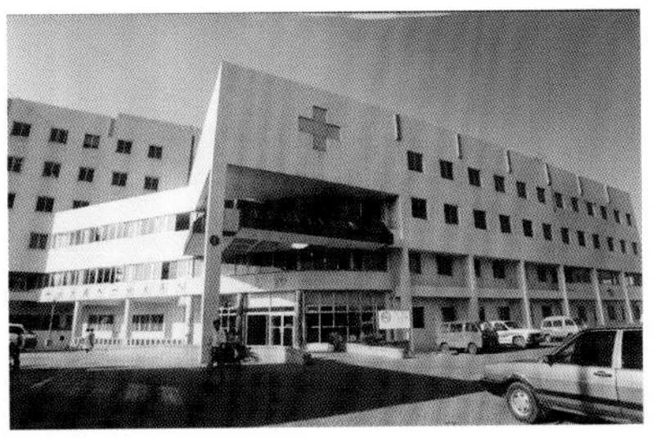

苏州大学附属第一医院
Soochow University Affiliated First Hospital

项目类型：医疗建筑
项目地点：江苏苏州
竣工时间：1994
Project Type：Medical building
Location：Suzhou, Jiangsu
Completion：1994

2001—2009 >>>

苏州图书馆
Suzhou Library

项目类型：图书馆
项目地点：江苏苏州
竣工时间：2001
Project Type：Library
Location：Suzhou, Jiangsu
Completion：2001

苏州会议中心二期
Suzhou Convention Center Phase II

项目类型：会务会展
项目地点：江苏苏州
竣工时间：2001
Project Type：Conference and exhibition
Location：Suzhou, Jiangsu
Completion：2001

泰州市中级人民法院审判办公综合楼
Trial Office Building of Taizhou Intermediate People's Court

项目类型：司法建筑
项目地点：江苏泰州
竣工时间：2001
Project Type：Judicial building
Location：Taizhou, Jiangsu
Completion：2001

苏州体育中心游泳馆
Fitness Center of Suzhou Sports Center

项目类型：体育建筑
项目地点：江苏苏州
竣工时间：2002
Project Type：Sports building
Location：Suzhou, Jiangsu
Completion：2002

苏州日报社
Suzhou Daily News Building

项目类型：办公
项目地点：江苏苏州
竣工时间：2002
Project Type：Office
Location：Suzhou, Jiangsu
Completion：2002

扬州鉴真学院
Yangzhou Jianzhen College

项目类型：宗教建筑
项目地点：江苏扬州
竣工时间：2007
Project Type：Religious architecture
Location：Yangzhou, Jiangsu
Completion：2007

扬州体育馆
Yangzhou Gymnasium

项目类型：体育建筑
项目地点：江苏扬州
竣工时间：2005
Project Type：Sports building
Location：Yangzhou, Jiangsu
Completion：2005

扬州国际会展中心
Yangzhou International Convention and Exhibition Center

项目类型：会务会展
项目地点：江苏扬州
竣工时间：2002
Project Type：Conference and exhibition
Location：Yangzhou, Jiangsu
Completion：2002

国际科技园 2 期
SISPARK Phase II

项目类型：办公
项目地点：江苏苏州
竣工时间：2003
Project Type：Office
Location：Suzhou, Jiangsu
Completion：2003

星海大厦
Xinghai Building

项目类型：办公
项目地点：江苏苏州
竣工时间：2004
Project Type: Office
Location: Suzhou, Jiangsu
Completion: 2004

中银惠龙大厦
Zhongyin Huilong Building

项目类型：办公
项目地点：江苏苏州
竣工时间：2006
Project Type: Office
Location: Suzhou, Jiangsu
Completion: 2006

师惠大厦
Shihui Building

项目类型：邻里中心
项目地点：江苏苏州
竣工时间：2004
Project Type: Neighborhood Center
Location: Suzhou, Jiangsu
Completion: 2004

建园大厦
Jianyuan Building

项目类型：办公
项目地点：江苏苏州
竣工时间：2004
Project Type: Office
Location: Suzhou, Jiangsu
Completion: 2004

苏州工业园区人民法院
People's Court of Suzhou Industrial Park

项目类型：司法建筑
项目地点：江苏苏州
竣工时间：2005
Project Type: Judicial building
Location: Suzhou, Jiangsu
Completion: 2005

苏州伊莎中心
Suzhou Yisha Center

项目类型：办公
项目地点：江苏苏州
竣工时间：2004
Project Type: Office
Location: Suzhou, Jiangsu
Completion: 2004

苏州汽车南站
Suzhou South Bus Station

项目类型：交通基础设施
项目地点：江苏苏州
竣工时间：2004
Project Type: Transport infrastructure
Location: Suzhou, Jiangsu
Completion: 2004

苏州沧浪少年宫
Suzhou Canglang Children's Palace

项目类型：教育建筑
项目地点：江苏苏州
竣工时间：2006
Project Type: Educational building
Location: Suzhou, Jiangsu
Completion: 2006

湖东邻里中心
Hudong Neighborhood Center

项目类型：邻里中心
项目地点：江苏苏州
竣工时间：2004
Project Type: Neighborhood Center
Location: Suzhou, Jiangsu
Completion: 2004

2001—2009 >>>

美罗商城
MATRO Mall

项目类型： 商业
项目地点： 江苏苏州
竣工时间： 2004
Project Type： Commercial
Location： Suzhou, Jiangsu
Completion： 2004

苏州绿宝广场
Suzhou Green Treasure Plaza

项目类型： 商业
项目地点： 江苏苏州
竣工时间： 2006
Project Type： Commercial
Location： Suzhou, Jiangsu
Completion： 2006

苏州平江万达广场
Suzhou Pingjiang Wanda Plaza

项目类型： 商业综合体
项目地点： 江苏苏州
竣工时间： 2008
Project Type： Commercial complex
Location： Suzhou, Jiangsu
Completion： 2008

苏州工业园区市场大厦
Suzhou Industrial Park Market Building

项目类型： 办公
项目地点： 江苏苏州
竣工时间： 2006
Project Type： Office
Location： Suzhou, Jiangsu
Completion： 2006

苏州监狱
Suzhou Prison

项目类型： 监狱建筑
项目地点： 江苏苏州
竣工时间： 2009
Project Type： Prison building
Location： Suzhou, Jiangsu
Completion： 2009

上海青浦区朱家角中学
Shanghai Qingpu Zhujiajiao Middle School

项目类型： 教育建筑
项目地点： 上海
竣工时间： 2004
Project Type： Educational building
Location： Shanghai
Completion： 2004

苏州研究生城综合楼
Suzhou Graduate City Complex Building

项目类型： 教育建筑
项目地点： 江苏苏州
竣工时间： 2003
Project Type： Educational building
Location： Suzhou, Jiangsu
Completion： 2003

苏州工艺美术职业技术学院
Suzhou Art &Design Technology Institute

项目类型： 教育建筑
项目地点： 江苏苏州
竣工时间： 2008
Project Type： Educational building
Location： Suzhou, Jiangsu
Completion： 2008

苏州太湖高尔夫会所
Suzhou Taihu Golf Club

项目类型： 酒店
项目地点： 江苏苏州
竣工时间： 2007
Project Type： Hotel
Location： Suzhou, Jiangsu
Completion： 2007

环秀晓筑养生度假村
Huanxiu Xiaozhu Health Resort

项目类型：酒店
项目地点：江苏苏州
竣工时间：2009
Project Type：Hotel
Location：Suzhou, Jiangsu
Completion：2009

湖滨楼
Lakeside Building

项目类型：餐厅
项目地点：江苏苏州
竣工时间：2006
Project Type：Restaurant
Location：Suzhou, Jiangsu
Completion：2006

苏州市母子医疗保健中心
Suzhou Maternal and Child Health Care Center

项目类型：医院
项目地点：江苏苏州
竣工时间：2003
Project Type：Hospital
Location：Suzhou, Jiangsu
Completion：2003

常熟市公安局办公综合楼
Changshu Public Security Bureau Office Complex

项目类型：行政办公楼
项目地点：江苏常熟
竣工时间：2006
Project Type：Administrative office building
Location：Changshu, Jiangsu
Completion：2006

都市花园五期
Orchare Manors Phase V

项目类型：住宅
项目地点：江苏苏州
竣工时间：2003
Project Type：Residence
Location：Suzhou, Jiangsu
Completion：2003

华润 · 平门府
China Resources Pingmenfu

项目类型：住宅
项目地点：江苏苏州
竣工时间：2009
Project Type：Residence
Location：Suzhou, Jiangsu
Completion：2009

佳能新工厂
Canon New Factory

项目类型：工业厂房
项目地点：江苏苏州
竣工时间：2001
Project Type：Industrial plant
Location：Suzhou, Jiangsu
Completion：2001

软件大厦
Software Building

项目类型：研发
项目地点：江苏苏州
竣工时间：2009
Project Type: Research and development
Location：Suzhou, Jiangsu
Completion：2009

毅嘉电子苏州有限公司
ICHIA Electronic Suzhou Co., Ltd

项目类型：工业厂房
项目地点：江苏苏州
竣工时间：2003
Project Type：Industrial plant
Location：Suzhou, Jiangsu
Completion：2003

星海街 9 号改造
Reconstruction of No. 9 Xinghai Street

项目类型：既有建筑改造 / 办公
项目地点：江苏苏州
项目规模：12673m²
Project Type：Office
Location：Suzhou, Jiangsu
Project scale：12673m²

西交利物浦大学科研楼
Research Building of Xijiao Liverpool University

项目类型：教育建筑
项目地点：江苏苏州
项目规模：45041m²
Project Type：Educational building
Location：Suzhou, Jiangsu
Project scale：45041m²

苏州太湖文化论坛国际会议中心
Taihu Culture Forum International Conference Center

项目类型：会议中心
项目地点：江苏苏州
项目规模：65783m²
Project Type：Conference center
Location：Suzhou, Jiangsu
Project scale：65783m²

书香世家 · 平江府
Scholarly Inn Pingjiang Mansion

项目类型：酒店
项目地点：江苏苏州
项目规模：19233m²
Project Type：Hotel
Location：Suzhou, Jiangsu
Project scale：19233m²

苏州高新区科技大厦
SND Science and Technology Building

项目类型：办公
项目地点：江苏苏州
项目规模：113167m²
Project Type：Office
Location：Suzhou, Jiangsu
Project scale：113167m²

吴中大厦
Wuzhong Building

项目类型：办公
项目地点：江苏苏州
项目规模：51287m²
Project Type：Office
Location：Suzhou, Jiangsu
Project scale：51287m²

苏尔寿泵业有限公司厂房和办公楼
Plant and Office Building of Sulzer Pump Co., Ltd

项目类型：工业厂房
项目地点：江苏苏州
项目规模：28030m²
Project Type：Industrial plant
Location：Suzhou, Jiangsu
Project scale：28030m²

苏州慈济园区
Suzhou Tzu Chi Park

项目类型：文化建筑
项目地点：江苏苏州
项目规模：63218m²
Project Type：Cultural Architecture
Location：Suzhou, Jiangsu
Project scale：63218m²

苏州科技城独立式标准厂房
MEDPARK

项目类型：科技研发
项目地点：江苏苏州
项目规模：11552m²
Project Type：Scientific Research
Location：Suzhou, Jiangsu
Project scale：11552m²

苏州东山宾馆一期绛云楼改造
Dongshan Hotel Phase I

项目类型：酒店
项目地点：江苏苏州
项目规模：2566m²
Project Type：Hotel
Location：Suzhou, Jiangsu
Project scale：2566m²

苏州工业园区招商银行大厦
China Merchants Bank Building

项目类型：金融办公
项目地点：江苏苏州
项目规模：44850m²
Project Type：Financial office
Location：Suzhou, Jiangsu
Project scale：44850m²

苏州广播电视总台新媒体中心
SBS New Media Center

项目类型：办公
项目地点：江苏苏州
项目规模：18260m²
Project Type：Office
Location：Suzhou, Jiangsu
Project scale：18260m²

苏州交通一号线工程控制中心大楼
Control Center Building of Suzhou Traffic Line 1

项目类型：办公
项目地点：江苏苏州
项目规模：57697m²
Project Type：Office
Location：Suzhou, Jiangsu
Project scale：57697m²

杨舍老街
Yangshe Old Street

项目类型：商业
项目地点：江苏苏州
项目规模：35050m²
Project Type：Commercial
Location：Suzhou, Jiangsu
Project scale：35050m²

中新置地月亮湾酒店
CSLAND Moon Bay Hotel

项目类型：酒店
项目地点：江苏苏州
项目规模：74124m²
Project Type：Hotel
Location：Suzhou, Jiangsu
Project scale：74124m²

星海生活广场
Xinghai Central Station

项目类型：地下空间
项目地点：江苏苏州
项目规模：52273m²
Project Type：Underground space
Location：Suzhou, Jiangsu
Project scale：52273m²

苏州汽车北站改建工程
Suzhou North Bus Station

项目类型：交通设施
项目地点：江苏苏州
项目规模：24778m²
Project Type：Transportation facilities
Location：Suzhou, Jiangsu
Project scale：24778m²

苏州吴中商务中心
Suzhou Wuzhong Business Center

项目类型：办公
项目地点：江苏苏州
项目规模：124224m²
Project Type：Office
Location：Suzhou, Jiangsu
Project scale：124224m²

苏州市沧浪实验中学
Suzhou Canglang Experimental Middle School

项目类型：教育建筑
项目地点：江苏苏州
项目规模：38082m²
Project Type：Educational building
Location：Suzhou, Jiangsu
Project scale：38082m²

江阴华士城市主题公园
Jiangyin Huashi City Theme Park

项目类型：公园景观
项目地点：江苏江阴
项目规模：186000m²
Project Type：Park landscape
Location：Jiangyin, Jiangsu
Project scale：186000m²

苏州高新区展示馆
SND Exhibition Hall

项目类型：文化场馆
项目地点：江苏苏州
项目规模：14022m²
Project Type：Cultural venues
Location：Suzhou, Jiangsu
Project scale：14022m²

苏州轨交一号线天平车辆段与综合基地
Suzhou Rail Transit Line 1 Tianping Depot and Comprehensive Base

项目类型：轨道交通
项目地点：江苏苏州
项目规模：80696m²
Project Type：Rail transit
Location：Suzhou, Jiangsu
Project scale：80696m²

沧浪新城社区服务中心
Canglang New City Community Service Center

项目类型：商业
项目地点：江苏苏州
项目规模：27097m²
Project Type：Commercial
Location：Suzhou, Jiangsu
Project scale：27097m²

苏州纳米科技城一期工程
Nano Polis Phase I Project

项目类型：科研办公
项目地点：江苏苏州
项目规模：107733m²
Project Type：Scientific research office
Location：Suzhou, Jiangsu
Project scale：107733m²

致远国际大厦
Zhiyuan International Business Building

项目类型：办公
项目地点：江苏苏州
项目规模：103323m²
Project Type：Office
Location：Suzhou, Jiangsu
Project scale：103323m²

苏州工业园区青剑湖学校
Qingjianhu School of Suzhou Industrial Park

项目类型：教育建筑
项目地点：江苏苏州
项目规模：47231m²
Project Type：Educational building
Location：Suzhou, Jiangsu
Project scale：47231m²

苏州信托大厦
Suzhou Trust Building

项目类型：办公
项目地点：江苏苏州
项目规模：69104m²
Project Type：Office
Location：Suzhou, Jiangsu
Project scale：69104m²

太仓万达广场南区（购物中心）
Taicang Wanda Plaza

项目类型：商业
项目地点：江苏苏州
项目规模：204336m²
Project Type：Commercial
Location：Suzhou, Jiangsu
Project scale：204336m²

尹山湖管理用房
Yinshan Lake Management Room

项目类型：市政基础设施
项目地点：江苏苏州
项目规模：11060m²
Project Type：Municipal infrastructure
Location：Suzhou, Jiangsu
Project scale：11060m²

吴中现代文体中心
Wuzhong Modern Culture and Sports Center

项目类型：文体 & 办公
项目地点：江苏苏州
项目规模：78023m²
Project Type：Culture and sports & office
Location：Suzhou, Jiangsu
Project scale：78023m²

潘祖荫故居改造（中路后半部及东路）
Reconstruction of Pan Zuyin's Former Residence

项目类型：酒店
项目地点：江苏苏州
项目规模：1800m²
Project Type：Hotel
Location：Suzhou, Jiangsu
Project scale：1800m²

苏州德威国际学校（一期）
Dulwich College Suzhou

项目类型：教育建筑
项目地点：江苏苏州
项目规模：25749m²
Project Type：Educational building
Location：Suzhou, Jiangsu
Project scale：25749m²

苏州工业园区公积金管理中心大厦
SIP Provident Fund Management Center Building

项目类型：办公
项目地点：江苏苏州
项目规模：68440m²
Project Type：Office
Location：Suzhou, Jiangsu
Project scale：68440m²

苏州高铁站枢纽区综合开发项目
Suzhou High speed Railway Station Hub Comprehensive Development Project

项目类型：交通枢纽
项目地点：江苏苏州
项目规模：11552m²
Project Type：Transportation hub
Location：Suzhou, Jiangsu
Project scale：11552m²

黄金水岸中心酒店
Coast Resort

项目类型：酒店
项目地点：江苏苏州
项目规模：39744m²
Project Type：Hotel
Location：Suzhou, Jiangsu
Project scale：39744m²

DK20090256 号地块项目
DK20090256 Plot Project

项目类型：办公
项目地点：江苏苏州
项目规模：41778m²
Project Type：Office
Location：Suzhou, Jiangsu
Project scale：41778m²

苏州工业园区交通银行大厦
SIP Bank of Communications Building

项目类型：金融办公
项目地点：江苏苏州
项目规模：64468m²
Project Type：Financial office
Location：Suzhou, Jiangsu
Project scale：64468m²

苏州市公安（应急）指挥中心
Suzhou Police (Emergency) Command Center

项目类型：办公
项目地点：江苏苏州
项目规模：81102m²
Project Type：Office
Location：Suzhou, Jiangsu
Project scale：11552m²

苏州科技城人才配套服务基地一区
Suzhou science and technology Town Talent Supporting Service Base Zone I

项目类型：商业
项目地点：江苏苏州
项目规模：61578m²
Project Type：Commercial
Location：Suzhou, Jiangsu
Project scale：61578m²

426 地块办公大楼
426 Plot Office Building

项目类型：金融办公
项目地点：江苏苏州
项目规模：48529m²
Project Type：Financial office
Location：Suzhou, Jiangsu
Project scale：48529m²

中银大厦
Bank of China Building

项目类型：金融办公
项目地点：江苏苏州
项目规模：99640m²
Project Type：Financial office
Location：Suzhou, Jiangsu
Project scale：99640m²

独墅湖高教区西交利物浦大学行政信息楼
Xijiao Liverpool University Administrative Information Building

项目类型：教育建筑
项目地点：江苏苏州
项目规模：59922m²
Project Type：Educational building
Location：Suzhou, Jiangsu
Project scale：59922m²

江苏移动苏州分公司工业园区新综合大楼
New Complex Building of Jiangsu Mobile Suzhou Branch Industrial Park

项目类型：办公
项目地点：江苏苏州
项目规模：85696m²
Project Type：Office
Location：Suzhou, Jiangsu
Project scale：85696m²

苏州工业园区恒宇国际中心二期
SIP Hengyu International Center Phase II

项目类型：商业综合体
项目地点：江苏苏州
项目规模：55996m^2
Project Type：Commercial complex
Location：Suzhou, Jiangsu
Project scale：55996m^2

苏州港华燃气研发大楼
Suzhou Ganghua Gas R&D Building

项目类型：办公
项目地点：江苏苏州
项目规模：70689m^2
Project Type：Office
Location：Suzhou, Jiangsu
Project scale：70689m^2

金阊人力资源服务大厦
Jinchang Human Resources Service Building

项目类型：办公
项目地点：江苏苏州
项目规模：17364m^2
Project Type：Office
Location：Suzhou, Jiangsu
Project scale：17364m^2

苏州经贸职业技术学院三期工程
Suzhou Institute of Trade&Commerce Phase III Project

项目类型：教育建筑
项目地点：江苏苏州
项目规模：40944m^2
Project Type：Educational building
Location：Suzhou, Jiangsu
Project scale：40944m^2

吴江总部经济大楼
Wujiang Headquarters Economic Building

项目类型：办公
项目地点：江苏苏州
项目规模：69533m^2
Project Type：Office
Location：Suzhou, Jiangsu
Project scale：69533m^2

苏州老年公寓（颐养家园）项目
Suzhou Elderly Apartment Project

项目类型：老年公寓
项目地点：江苏苏州
项目规模：81592m^2
Project Type：Senior apartment
Location：Suzhou, Jiangsu
Project scale：81592m^2

张省艺术馆
Zhangxing Art Museum

项目类型：文化场馆
项目地点：江苏苏州
项目规模：3072m^2
Project Type：Cultural venues
Location：Suzhou, Jiangsu
Project scale：3072m^2

斜塘老街四期
Xietang Old Street Phase IV

项目类型：商业街区
项目地点：江苏苏州
项目规模：27768m^2
Project Type：Commercial street
Location：Suzhou, Jiangsu
Project scale：27768m^2

天域大厦
Tianyu Building

项目类型：办公
项目地点：江苏苏州
项目规模：61794m^2
Project Type：Office
Location：Suzhou, Jiangsu
Project scale：61794m^2

第九届江苏省园艺博览会园博园工程 B 馆
Hall B of the 9th Jiangsu Horticultural Exposition Park Expo Project

项目类型：文化建筑
项目地点：江苏苏州
项目规模：24520m^2
Project Type：Cultural architecture
Location：Suzhou, Jiangsu
Project scale：24520m^2

江苏银行苏州分行园区办公大楼
Bank of Jiangsu Suzhou Branch

项目类型：金融办公
项目地点：江苏苏州
项目规模：48588m^2
Project Type：Financial office
Location：Suzhou, Jiangsu
Project scale：48588m^2

苏州妇女儿童活动中心迁建工程
Suzhou Women and Children's Activity Center

项目类型：文体场馆
项目地点：江苏苏州
项目规模：44146m^2
Project Type：Culture and sports venues
Location：Suzhou, Jiangsu
Project scale：44146m^2

苏州市吴江区笠泽实验初级中学
Lize Experimental Junior High School

项目类型：教育建筑
项目地点：江苏苏州
项目规模：35952m^2
Project Type：Educational building
Location：Suzhou, Jiangsu
Project scale：35952m^2

月亮湾地块 B05 项目
Moon Bay Plot B05 Project

项目类型：超高层
项目地点：江苏苏州
项目规模：136574m^2
Project Type：Super high-rise
Location：Suzhou, Jiangsu
Project scale：136574m^2

宝时得中国总部（一期）
POSTIEC China Headquarters (Phase I)

项目类型：研发办公
项目地点：江苏苏州
项目规模：33304m^2
Project Type：R&D office
Location：Suzhou, Jiangsu
Project scale：33304m^2

国发 · 平江大厦
Guofa Pingjiang Building

项目类型：办公
项目地点：江苏苏州
项目规模：115533m²
Project Type：Office
Location：Suzhou, Jiangsu
Project scale：115533m²

甪直游客服务中心
Luzhi Tourist Service Center

项目类型：游客中心
项目地点：江苏苏州
项目规模：14285m²
Project Type：Tourist center
Location：Suzhou, Jiangsu
Project scale：14285m²

国库支付中心
Treasury Payment Center

项目类型：办公
项目地点：江苏苏州
项目规模：60741m²
Project Type：Office
Location：Suzhou, Jiangsu
Project scale：60741m²

吴中区公检楼
Wuzhong District Public Inspection Building

项目类型：办公
项目地点：江苏苏州
项目规模：67799m²
Project Type：Office
Location：Suzhou, Jiangsu
Project scale：67799m²

锦峰大厦
Jinfeng Building

项目类型：商业办公
项目地点：江苏苏州
项目规模：193058m²
Project Type：Commercial and office
Location：Suzhou, Jiangsu
Project scale：193058m²

苏悦广场
Suyue Plaza

项目类型：商业办公
项目地点：江苏苏州
项目规模：221575m²
Project Type：Commercial and office
Location：Suzhou, Jiangsu
Project scale：221575m²

苏州高新区永旺梦乐城
SND Aeon

项目类型：商业
项目地点：江苏苏州
项目规模：163782m²
Project Type：Commercial
Location：Suzhou, Jiangsu
Project scale：163782m²

宿迁市宿城区实验小学迁建工程
Sucheng Experimental Primary School

项目类型：教育建筑
项目地点：江苏宿迁
项目规模：38823m²
Project Type：Educational building
Location：Suqian, Jiangsu
Project scale：38823m²

顾村规划展示馆
Gucun Planning Exhibition Hall

项目类型：文化场馆
项目地点：上海
项目规模：13321m²
Project Type：Cultural venues
Location：Shanghai
Project scale：13321m²

东山宾馆叠翠楼
Diecuilou, Dongshan Hotel

项目类型：酒店
项目地点：江苏苏州
项目规模：24891m²
Project Type：Hotel
Location：Suzhou, Jiangsu
Project scale：24891m²

苏地 2009-B-76 号地块综合楼
Sudi 2009-B-76 Plot Complex Building

项目类型：商业办公
项目地点：江苏苏州
项目规模：115544m²
Project Type：Commercial and office
Location：Suzhou, Jiangsu
Project scale：115544m²

苏州太湖科技产业园科技研发大楼项目
Suzhou Taihu Science and Technology Industrial Park R&D Building Project

项目类型：办公
项目地点：江苏苏州
项目规模：40696m²
Project Type：Office
Location：Suzhou, Jiangsu
Project scale：40696m²

苏州工业园区久龄公寓
Jiuling Apartment

项目类型：老年公寓
项目地点：江苏苏州
项目规模：60158m²
Project Type：Senior apartment
Location：Suzhou, Jiangsu
Project scale：60158m²

中科院地理信息与文化科技产业基地
Geographic Information and Cultural Technology Industry Base of Chinese Academy of Sciences

项目类型：办公
项目地点：江苏苏州
项目规模：39298m²
Project Type：Office
Location：Suzhou, Jiangsu
Project scale：39298m²

苏州独墅湖地块项目 51 号楼
Suzhou Dushu Lake Plot Project Building 51

项目类型：商业配套
项目地点：江苏苏州
项目规模：2401m²
Project Type：Commercial
Location：Suzhou, Jiangsu
Project scale：2401m²

苏州中心广场项目 A 地块、中轴线
Suzhou Center

项目类型：商业综合体
项目地点：江苏苏州
项目规模：256156m^2
Project Type：Commercial complex
Location：Suzhou, Jiangsu
Project scale：256156m^2

龙湖 · 苏地 2013-G-18 号地块（四号地块）项目
LONGFOR Shishan Tianjie

项目类型：商业综合体
项目地点：江苏苏州
项目规模：278062m^2
Project Type：Commercial complex
Location：Suzhou, Jiangsu
Project scale：278062m^2

吴江中学初中部
Middle School Department of Wujiang High School

项目类型：教育建筑
项目地点：江苏苏州
项目规模：29140m^2
Project Type：Educational building
Location：Suzhou, Jiangsu
Project scale：29140m^2

苏州工业园区星汇学校
SIP Xinghui School

项目类型：教育建筑
项目地点：江苏苏州
项目规模：65911m^2
Project Type：Educational building
Location：Suzhou, Jiangsu
Project scale：65911m^2

苏州工业园区淞泽小学
SIP Songze School

项目类型：教育建筑
项目地点：江苏苏州
项目规模：27363m^2
Project Type：Educational building
Location：Suzhou, Jiangsu
Project scale：27363m^2

苏滁产业园国际商务中心
International Business Center of Suchu Modern Industrial Park

项目类型：办公
项目地点：安徽滁州
项目规模：50507m^2
Project Type：Office
Location：Chuzhou, Anhui
Project scale：50507m^2

中国医药城（泰州）会展交易中心二期
China Pharmaceutical City (Taizhou) Exhibition and Trading Center Phase II

项目类型：会展中心
项目地点：江苏泰州
项目规模：105895m^2
Project Type：Convention and exhibition center
Location：Taizhou, Jiangsu
Project scale：105895m^2

姑苏软件园项目
Gusu Software Park Project

项目类型：办公
项目地点：江苏苏州
项目规模：204251m^2
Project Type：Office
Location：Suzhou, Jiangsu
Project scale：204251m^2

西交利物浦大学南校区规划及一期工程
South Campus of Xi'an Jiaotong Liverpool University

项目类型：教育建筑
项目地点：江苏苏州
项目规模：91834m^2
Project Type：Educational building
Location：Suzhou, Jiangsu
Project scale：91834m^2

顾村菊泉文化展示馆
Gucun Juquan Culture Exhibition Hall

项目类型：文化场馆
项目地点：上海
项目规模：8144m^2
Project Type：Cultural venues
Location：Shanghai
Project scale：8144m^2

苏州吴中凤凰广场
Wuzhong Phoenix Plaza

项目类型：商业办公
项目地点：江苏苏州
项目规模：31790m^2
Project Type：Commercial office
Location：Suzhou, Jiangsu
Project scale：31790m^2

苏州高新区人民医院二期工程项目
SND People's Hospital Phase II Project

项目类型：医院
项目地点：江苏苏州
项目规模：95088m^2
Project Type：Hospital
Location：Suzhou, Jiangsu
Project scale：95088m^2

千鹤湾老年公寓
Qianhewan Elderly Apartment

项目类型：老年公寓
项目地点：江苏盐城
项目规模：112347m^2
Project Type：Senior apartment
Location：Yancheng, Jiangsu
Project scale：112347m^2

东太湖防汛物资仓库工程
East Taihu Lake Flood Control Material Warehouse

项目类型：市政基础设施
项目地点：江苏苏州
项目规模：1379m^2
Project Type：Municipal infrastructure
Location：Suzhou, Jiangsu
Project scale：1379m^2

苏州丰隆城市中心
Suzhou Fenglong City Center

项目类型：商业综合体
项目地点：江苏苏州
项目规模：410987m^2
Project Type：Commercial complex
Location：Suzhou, Jiangsu
Project scale：410987m^2

苏州高新区实验初级中学东校区扩建工程
SND Experimental Junior Middle School East Campus Expansion Project

项目类型：教育建筑
项目地点：江苏苏州
项目规模：18326m^2
Project Type：Educational building
Location：Suzhou, Jiangsu
Project scale：18326m^2

建屋广场 C 座
Building C, GENWAY Square

项目类型：办公
项目地点：江苏苏州
项目规模：76764m^2
Project Type：Office
Location：Suzhou, Jiangsu
Project scale：76764m^2

江苏省苏州实验中学科技城校
Suzhou Experimental Middle School Science City School

项目类型：教育建筑
项目地点：江苏苏州
项目规模：90362m^2
Project Type：Educational building
Location：Suzhou, Jiangsu
Project scale：90362m^2

苏州系统医学研究所新建项目（一期）
New Project of Suzhou Institute of Systemic Medicine (Phase I)

项目类型：科研办公
项目地点：江苏苏州
项目规模：45136m^2
Project Type：Scientific research office
Location：Suzhou, Jiangsu
Project scale：45136m^2

苏州科技城第二实验小学
Suzhou Science and Technology Town No. 2 Experimental Primary School

项目类型：教育建筑
项目地点：江苏苏州
项目规模：36379m^2
Project Type：Educational building
Location：Suzhou, Jiangsu
Project scale：36379m^2

苏州太湖万丽万豪酒店
Suzhou Taihu Wanli Marriott Hotel

项目类型：酒店
项目地点：江苏苏州
项目规模：118408m^2
Project Type：Hotel
Location：Suzhou, Jiangsu
Project scale：118408m^2

南溪江商务中心
Nanxijiang Business Center

项目类型：商业
项目地点：江苏苏州
项目规模：120585m^2
Project Type：Commercial
Location：Suzhou, Jiangsu
Project scale：120585m^2

通园路停保场
Tongyuan Road Parking and Maintenance Yard

项目类型：市政基础设施
项目地点：江苏苏州
项目规模：51123m^2
Project Type：Municipal infrastructure
Location：Suzhou, Jiangsu
Project scale：51123m^2

津西新天地
Jinxi New Tiandi

项目类型：商业综合体
项目地点：江苏苏州
项目规模：64347m^2
Project Type：Commercial complex
Location：Suzhou, Jiangsu
Project scale：64347m^2

苏州阳光城翡丽湾
Suzhou Sunshine City Feiliwan

项目类型：教育建筑
项目地点：江苏苏州
项目规模：11552m^2
Project Type：Educational building
Location：Suzhou, Jiangsu
Project scale：11552m^2

雅致 · 湖沁阁
Mehood Elegant Hotel

项目类型：酒店
项目地点：江苏苏州
项目规模：27768m^2
Project Type：Hotel
Location：Suzhou, Jiangsu
Project scale：27768m^2

昆山虹祺路商业街
Kunshan Hongqi Road Business Street

项目类型：商业
项目地点：江苏昆山
项目规模：80022m^2
Project Type：Commercial
Location：Kunshan, Jiangsu
Project scale：80022m^2

冯梦龙纪念馆工程
Feng Menglong Memorial Hall Project

项目类型：文化建筑
项目地点：江苏苏州
项目规模：394m^2
Project Type：Cultural architecture
Location：Suzhou, Jiangsu
Project scale：394m^2

晶桥云鹤山村综合服务中心
Jingqiao Yunheshan Village Comprehensive Service Center

项目类型：村镇建筑
项目地点：江苏南京
项目规模：845m^2
Project Type：Village architecture
Location：Nanjing, Jiangsu
Project scale：845m^2

江苏硅谷展示馆
Jiangsu Silicon Valley Exhibition Hall

项目类型：办公
项目地点：江苏句容
项目规模：4461m^2
Project Type：Office
Location：Jurong, Jiangsu
Project scale：4461m^2

文旅万和广场
Wenlv Wanhe Square

项目类型：商业办公
项目地点：江苏苏州
项目规模：81707m^2
Project Type：Commercial and office
Location：Suzhou, Jiangsu
Project scale：81707m^2

大兆瓦风机新园区项目
Megawatt Wind Turbine New Park Project

项目类型：工业建筑
项目地点：江苏无锡
项目规模：44014m^2
Project Type：Industrial building
Location：Wuxi, Jiangsu
Project scale：44014m^2

中国移动苏州研发中心项目二期
China Mobile Suzhou R&D Center Project Phase II

项目类型：办公
项目地点：江苏苏州
项目规模：69510m^2
Project Type：Office
Location：Suzhou, Jiangsu
Project scale：69510m^2

吴郡幼儿园
Wujun Kindergarten

项目类型：教育建筑
项目地点：江苏苏州
项目规模：20266m^2
Project Type：Educational building
Location：Suzhou, Jiangsu
Project scale：20266m^2

苏地 2016-WG-10 号地块 1 号楼
Building 1 of Sudi 2016-WG-10 Plot

项目类型：商业
项目地点：江苏苏州
项目规模：3327m^2
Project Type：Commercial
Location：Suzhou, Jiangsu
Project scale：3327m^2

国裕大厦二期
Guoyu Building Phase II

项目类型：办公
项目地点：江苏苏州
项目规模：52290m^2
Project Type：Office
Location：Suzhou, Jiangsu
Project scale：52290m^2

新湖广场
Xinhu Square

项目类型：商业
项目地点：江苏苏州
项目规模：43473m^2
Project Type：Commercial
Location：Suzhou, Jiangsu
Project scale：43473m^2

高铁新城体育馆项目
High speed railway new city stadium project

项目类型：文体建筑
项目地点：江苏苏州
项目规模：9542m^2
Project Type：Culture & sports building
Location：Suzhou, Jiangsu
Project scale：9542m^2

苏州港口发展大厦
Suzhou Port Development Building

项目类型：商业办公
项目地点：江苏苏州
项目规模：94231m^2
Project Type：Commercial and office
Location：Suzhou, Jiangsu
Project scale：94231m^2

微创骨科苏州项目 -2 号综合楼
Microinvasive Orthopedics Suzhou Project 2 # Complex Building

项目类型：科技研发
项目地点：江苏苏州
项目规模：8043m^2
Project Type：Scientific research
Location：Suzhou, Jiangsu
Project scale：8043m^2

苏州农业职业技术学院东山分校
Suzhou Polytechnic Institute of Agriculture Dongshan Campus

项目类型：教育建筑
项目地点：江苏苏州
项目规模：33847m^2
Project Type：Educational building
Location：Suzhou, Jiangsu
Project scale：33847m^2

环古城南新路地块改造整治工程
Reconstruction and Renovation Project of Huangucheng South New Road Plot

项目类型：商业
项目地点：江苏苏州
项目规模：21161m^2
Project Type：Commercial
Location：Suzhou, Jiangsu
Project scale：21161m^2

丰华国际服务中心
Fenghua International Service Center

项目类型：办公
项目地点：江苏苏州
项目规模：112987m^2
Project Type：Office
Location：Suzhou, Jiangsu
Project scale：112987m^2

苏州工业园区青剑湖高中
Qingjianhu High School in SIP

项目类型：建筑
项目地点：江苏苏州
项目规模：90384m^2
Project Type：Educational building
Location：Suzhou, Jiangsu
Project scale：90384m^2

常熟永旺梦乐城
Changshu AEON

项目类型：商业
项目地点：江苏常熟
项目规模：180898m^2
Project Type：Commercial
Location：Changshu, Jiangsu
Project scale：180898m^2

玉山幼儿园
Yushan Kindergarten

项目类型：教育建筑
项目地点：江苏苏州
项目规模：10669m²
Project Type：Educational building
Location：Suzhou, Jiangsu
Project scale：10669m²

城铁新城幼儿园
Chengtie Xincheng Kindergarten

项目类型：教育建筑
项目地点：江苏苏州
项目规模：11543m²
Project Type：Educational building
Location：Suzhou, Jiangsu
Project scale：11543m²

中国科学院空天信息创新研究院苏州园区
Suzhou Park Institute of Electrics, Chinese Academy of Sciences

项目类型：办公
项目地点：江苏苏州
项目规模：47266m²
Project Type：Office
Location：Suzhou, Jiangsu
Project scale：47266m²

朗高电机新能源汽车电机工厂
Suzhou LEGO New Energy Automobile Motor Factory

项目类型：工业研发
项目地点：江苏苏州
项目规模：26865m²
Project Type：Industrial R&D
Location：Suzhou, Jiangsu
Project scale：26865m²

XDG-2014-39 号地块开发建设项目
XDG-2014-39 Plot Project

项目类型：酒店
项目地点：江苏无锡
项目规模：57000m²
Project Type：Hotel
Location：Wuxi, Jiangsu
Project scale：57000m²

冯梦龙村党建文化馆二期工程
Fengmenglong Village Cultural Center Phase II

项目类型：文化建筑
项目地点：江苏苏州
项目规模：258m²
Project Type：Cultural architecture
Location：Suzhou, Jiangsu
Project scale：258m²

昆山加拿大国际学校二期
Kunshan Canada International School Phase II

项目类型：教育建筑
项目地点：江苏昆山
项目规模：8458m²
Project Type：Educational building
Location：Kunshan, Jiangsu
Project scale：8458m²

苏州太美逸郡酒店
G-Luxe By Gloria Suzhou Taimei

项目类型：酒店
项目地点：江苏苏州
项目规模：20935m²
Project Type：Hotel
Location：Suzhou, Jiangsu
Project scale：20935m²

苏州高新区滨河实验小学校
SND Binhe Experimental Primary School

项目类型：教育建筑
项目地点：江苏苏州
项目规模：49540m²
Project Type：Educational building
Location：Suzhou, Jiangsu
Project scale：49540m²

枫桥工业园改造一期
Fengqiao Industrial Park Reconstruction Phase I

项目类型：工业研发
项目地点：江苏苏州
项目规模：57407m²
Project Type：Industrial R&D
Location：Suzhou, Jiangsu
Project scale：57407m²

苏州工业园区钟南街义务制学校
SIP Zhongnan Street Compulsory School

项目类型：教育建筑
项目地点：江苏苏州
项目规模：80952m²
Project Type：Educational building
Location：Suzhou, Jiangsu
Project scale：80952m²

绿景 NEO 大厦
LGEM NEO Building

项目类型：商业办公
项目地点：江苏苏州
项目规模：81581m²
Project Type：Commercial and office
Location：Suzhou, Jiangsu
Project scale：81581m²

潘祖荫故居三期
Pan Zuyin's Former Residence Phase III

项目类型：城市更新
项目地点：江苏苏州
项目规模：861m²
Project Type：Urban renewal
Location：Suzhou, Jiangsu
Project scale：861m²

常熟市高新区三环小学及幼儿园工程
Sanhuan Primary School and Kindergarten Project

项目类型：教育建筑
项目地点：江苏苏州
项目规模：48293m²
Project Type：Educational building
Location：Suzhou, Jiangsu
Project scale：48293m²

苏州高新区马舍山酒店改扩建项目
Naked Warter

项目类型：酒店
项目地点：江苏苏州
项目规模：13760m²
Project Type：Hotel
Location：Suzhou, Jiangsu
Project scale：13760m²

中铁第四勘察设计院苏州创意产业园
Suzhou Creative Industrial Park of China Railway Fourth Survey and Design Institute

项目类型：办公
项目地点：江苏苏州
项目规模：32862m^2
Project Type：Office
Location：Suzhou, Jiangsu
Project scale：32862m^2

吴江太湖新城软件园综合大楼
Comprehensive Building of Software Park in Wujiang Taihu New Town

项目类型：办公
项目地点：江苏苏州
项目规模：42014m^2
Project Type：Office
Location：Suzhou, Jiangsu
Project scale：42014m^2

苏州阳澄喜柯大酒店
Suzhou Yangcheng Siko Grang Hotel

项目类型：酒店
项目地点：江苏苏州
项目规模：42958m^2
Project Type：Hotel
Location：Suzhou, Jiangsu
Project scale：42958m^2

新浒幼儿园
Xinxu Kindergarten

项目类型：教育建筑
项目地点：江苏苏州
项目规模：16594m^2
Project Type：Educational building
Location：Suzhou, Jiangsu
Project scale：16594m^2

西园养老护理院
Xiyuan Nursing Home

项目类型：养老建筑
项目地点：江苏苏州
项目规模：28256m^2
Project Type：Elderly care building
Location：Suzhou, Jiangsu
Project scale：28256m^2

苏州湾实验初级中学艺体馆
Suzhou Bay Experimental Junior Middle School Art and Sports Hall

项目类型：教育建筑
项目地点：江苏苏州
项目规模：6409m^2
Project Type：Educational building
Location：Suzhou, Jiangsu
Project scale：6409m^2

冯梦龙村山歌文化馆
Fengmenglong Village Folk Song Cultural Museum

项目类型：文化建筑
项目地点：江苏苏州
项目规模：2135m^2
Project Type：Cultural architecture
Location：Suzhou, Jiangsu
Project scale：2135m^2

树山村改造提升工程
Shushan Village Reconstruction and Upgrading Project

项目类型：乡村振兴
项目地点：江苏苏州
项目规模：103058m^2
Project Type：Rural vitalization
Location：Suzhou, Jiangsu
Project scale：103058m^2

苏州工业园区北部文体中心
SIP Northern Culture and Sports Center

项目类型：文体中心
项目地点：江苏苏州
项目规模：48848m^2
Project Type：Culture and sports center
Location：Suzhou, Jiangsu
Project scale：48848m^2

枫桥工业园改造二期
Fengqiao Industrial Park Reconstruction Phase II

项目类型：工业研发
项目地点：江苏苏州
项目规模：83678m^2
Project Type：Industrial R&D
Location：Suzhou, Jiangsu
Project scale：83678m^2

苏苑高级中学
Suyuan Haigh School

项目类型：教育建筑
项目地点：江苏苏州
项目规模：85878m^2
Project Type：Educational building
Location：Suzhou, Jiangsu
Project scale：85878m^2

梁丰初中西校区
Liangfeng Middle School West Campus

项目类型：教育建筑
项目地点：江苏苏州
项目规模：49712m^2
Project Type：Educational building
Location：Suzhou, Jiangsu
Project scale：49712m^2

苏州轨交四号线支线溪霞路站配套地下空间
Suzhou Taihu New Town Underground Space

项目类型：地下空间
项目地点：江苏苏州
项目规模：148990m^2
Project Type：Underground space
Location：Suzhou, Jiangsu
Project scale：148990m^2

芯汇湖大厦
Xinhuihu Building

项目类型：办公
项目地点：江苏苏州
项目规模：64894m^2
Project Type：Office
Location：Suzhou, Jiangsu
Project scale：64894m^2

石湖景区南石湖越来溪
South Shi Lake Yuelaixi of Shi Lake Scenic Area

项目类型：商业
项目地点：江苏苏州
项目规模：44797m^2
Project Type：Commercial
Location：Suzhou, Jiangsu
Project scale：44797m^2

苏州杵山交通枢纽
Suzhou Chushan Traffic Hub

项目类型：交通枢纽
项目地点：江苏苏州
项目规模：34253m^2
Project Type：Traffic Hub
Location：Suzhou, Jiangsu
Project scale：34253m^2

元和活力岛城市副中心提升改造工程设计
Design of Upgrading and Reconstruction Project of Yuanhe Vitality Island Sub Center

项目类型：商业中心
项目地点：江苏苏州
项目规模：11552m^2
Project Type：Commercial
Location：Suzhou, Jiangsu
Project scale：11552m^2

数据中心机房项目
Data Center Machine Room Project

项目类型：工业建筑
项目地点：江苏苏州
项目规模：14536m^2
Project Type：Industrial building
Location：Suzhou, Jiangsu
Project scale：14536m^2

昆山市公共卫生中心
Kunshan Public Health Center

项目类型：医疗建筑
项目地点：江苏昆山
项目规模：72428m^2
Project Type：Medical building
Location：Kunshan, Jiangsu
Project scale：72428m^2

江苏悦阳新建高标准厂房项目
Jiangsu Yueyang New High standard Factory Building Project

项目类型：工业建筑
项目地点：江苏盐城
项目规模：109456m^2
Project Type：Industrial building
Location：Yancheng, Jiangsu
Project scale：109456m^2

苏州市轨道交通线网档案中心
Suzhou Rail Transit Network Archives Center

项目类型：办公建筑
项目地点：江苏苏州
项目规模：32017m^2
Project Type：Office
Location：Suzhou, Jiangsu
Project scale：32017m^2

星海实验中学
Xinghai Experimental Middle School

项目类型：教育建筑
项目地点：江苏苏州
项目规模：88560m^2
Project Type：Educational building
Location：Suzhou, Jiangsu
Project scale：88560m^2

苏州乐园森林世界－动物托邦项目
Suzhou Paradise Forest World

项目类型：商业休闲建筑
项目地点：江苏苏州
项目规模：81751m^2
Project Type：Commercial and leisure building
Location：Suzhou, Jiangsu
Project scale：81751m^2

重庆哈罗国际学校
Harrow Chongqing

项目类型：教育建筑
项目地点：重庆
项目规模：61734m^2
Project Type：Educational building
Location：Chongqing
Project scale：61734m^2

吴郡小学
Wujun Primary School

项目类型：教育建筑
项目地点：江苏苏州
项目规模：35078m^2
Project Type：Educational building
Location：Suzhou, Jiangsu
Project scale：35078m^2

苏州绿岸锦绣澜山 9 号地块
Suzhou Green Bank Jinxiu Lanshan Plot 9

项目类型：商业建筑
项目地点：江苏苏州
项目规模：36437m^2
Project Type：Commercial
Location：Suzhou, Jiangsu
Project scale：36437m^2

行政中心 5 号楼维修改造设计
Maintenance and Reconstruction Design of Administrative Center Building 5

项目类型：行政办公
项目地点：江苏苏州
项目规模：35453m^2
Project Type：Administrative Office
Location：Suzhou, Jiangsu
Project scale：35453m^2

苏州工业园区人力资源服务产业园改造项目
SIP Human Resources Service Industrial Park Reconstruction Project

项目类型：城市更新
项目地点：江苏苏州
项目规模：47277m^2
Project Type：Urban renewal
Location：Suzhou, Jiangsu
Project scale：47277m^2

常熟市声谷展示中心
Changshu Acoustics Valley Exhibition Center

项目类型：城市更新
项目地点：江苏常熟
项目规模：2210m^2
Project Type：Urban renewal
Location：Changshu, Jiangsu
Project scale：2210m^2

南京天隆寺
Nanjing Tianlong Temple

项目类型：宗教建筑
项目地点：江苏南京
项目规模：44384m^2
Project Type：Religious architecture
Location：Nanjing, Jiangsu
Project scale：44384m^2

DK20170082 地块项目
DK20170082 Plot Project

项目类型：工业建筑
项目地点：江苏苏州
项目规模：89196m^2
Project Type：Industrial building
Location：Suzhou, Jiangsu
Project scale：89196m^2

浙商银行苏州分行
Zheshang Bank Suzhou Branch

项目类型：办公建筑
项目地点：江苏苏州
项目规模：42718m²
Project Type：Office
Location：Suzhou, Jiangsu
Project scale：42718m²

苏州市会议中心综合提升工程
Suzhou Conference Center Comprehensive Improvement Project

项目类型：城市更新
项目地点：江苏苏州
项目规模：87720m²
Project Type：Urban renewal
Location：Suzhou, Jiangsu
Project scale：87720m²

苏州评弹公园
Suzhou Balled Singing Park

项目类型：文化建筑
项目地点：江苏苏州
项目规模：5843m²
Project Type：Cultural architecture
Location：Suzhou, Jiangsu
Project scale：5843m²

苏州金普顿竹辉酒店
Suzhou Kimpton Bamboo Grove Hotel

项目类型：酒店建筑
项目地点：江苏苏州
项目规模：60457m²
Project Type：Hotel
Location：Suzhou, Jiangsu
Project scale：60457m²

随州大洪山落湖片区洪山寺游客服务中心
Hongshan Temple Tourist Service Center in the Luohu District of Dahongshan, Suizhou

项目类型：旅游建筑
项目地点：湖北随州
项目规模：87720m²
Project Type：Tourism building
Location：Suizhou, Hubei
Project scale：87720m²

苏科外附属幼儿园（玉屏路校区）
Sukewai Affiliated Kindergarten (Yuping Road Campus)

项目类型：教育建筑
项目地点：江苏苏州
项目规模：15468m²
Project Type：Educational building
Location：Suzhou, Jiangsu
Project scale：15468m²

苏州外国语学校吴中校区
Suzhou Foreign Language School Wuzhong Campus

项目类型：教育建筑
项目地点：江苏苏州
项目规模：98188m²
Project Type：Educational building
Location：Suzhou, Jiangsu
Project scale：98188m²

苏州荟同国际学校
Suzhou Huitong International School

项目类型：教育建筑
项目地点：江苏苏州
项目规模：121250m²
Project Type：Educational building
Location：Suzhou, Jiangsu
Project scale：121250m²

共创方程式科技研发服务中心
Gongchuang Formula Technology R&D Service Center

项目类型：办公建筑
项目地点：江苏苏州
项目规模：42825m²
Project Type：Office
Location：Suzhou, Jiangsu
Project scale：42825m²

苏州中盟科技项目
Suzhou Zhongmeng Technology Project

项目类型：办公建筑
项目地点：江苏苏州
项目规模：41930m²
Project Type：Office
Location：Suzhou, Jiangsu
Project scale：41930m²

骊住科技新工厂
LIXIL New Factory

项目类型：工业建筑
项目地点：江苏苏州
项目规模：46000m²
Project Type：Industrial building
Location：Suzhou, Jiangsu
Project scale：46000m²

苏州凯博易控产业园
Suzhou Ekontrol Industrial Park

项目类型：工业建筑
项目地点：江苏苏州
项目规模：125432m²
Project Type：Industrial building
Location：Suzhou, Jiangsu
Project scale：125432m²

苏州新光维医疗科技有限公司一期
Suzhou Scivita Medical Phase I

项目类型：工业建筑
项目地点：江苏苏州
项目规模：16805m²
Project Type：Industrial building
Location：Suzhou, Jiangsu
Project scale：16805m²

江苏经纬轨道交通设备有限公司新建厂房
Jiangsu Kingway Rail Transit Equipment New Factory Building

项目类型：工业建筑
项目地点：江苏苏州
项目规模：39178m²
Project Type：Industrial building
Location：Suzhou, Jiangsu
Project scale：39178m²

环元荡生态岸线提升工程
Ecological Shoreline Improvement Project Around Yuandang

项目类型：生态景观
项目地点：江苏苏州
项目规模：192000m²
Project Type：Ecological landscape
Location：Suzhou, Jiangsu
Project scale：192000m²

吴中水厂深度处理工程综合办公楼
Comprehensive Office Building for Deep Treatment Project of Wuzhong Water Plant

项目类型：办公建筑
项目地点：江苏苏州
项目规模：20973m^2
Project Type：Office
Location：Suzhou, Jiangsu
Project scale：20973m^2

和枫科创园设计
Hefeng Science and Technology Innovation Park

项目类型：工业建筑
项目地点 v：江苏苏州
项目规模：231821m^2
Project Type：Industrial building
Location：Suzhou, Jiangsu
Project scale：231821m^2

恒泰创新中心
Hengtai Innovation Center

项目类型：办公建筑
项目地点：江苏苏州
项目规模：93787m^2
Project Type：Office
Location：Suzhou, Jiangsu
Project scale：93787m^2

昆山东部医疗中心
Kunshan Eastern Medical Center

项目类型：医疗建筑
项目地点：江苏昆山
项目规模：249895m^2
Project Type：Medical building
Location：Kunshan, Jiangsu
Project scale：249895m^2

安元路幼儿园
Anyuan Road Kindergarten

项目类型：教育建筑
项目地点：江苏苏州
项目规模：15468m^2
Project Type：Educational building
Location：Suzhou, Jiangsu
Project scale：15468 m^2

江苏科瑞恩自动化科技有限公司新建厂房
Jiangsu Cowain New Factory Building

项目类型：工业建筑
项目地点：江苏昆山
项目规模：41289m^2
Project Type：Industrial building
Location：Kunshan, Jiangsu
Project scale：41289m^2